Fisheries Statistical Methods

Theory and Practical Manual

THE AUTHORS

J. Rujan is an Assistant Professor in the Department of Fisheries Extension, Economics, and Statistics at the College of Fisheries Science, St. Devasahayam Institute of Fisheries Science and Technology, Midalam, Kanniyakumari dist., Tamil Nadu. He graduated with a Bachelor's degree in Fisheries Science from Dr. MGR Fisheries College and Research Institute, Thalainayeru, Tamil Nadu. He obtained his Master's degree in Fisheries Science from the ICAR-Central Institute of Fisheries Education, Mumbai. He has received Gold Medals in both undergraduate and postgraduate programs. Additionally, he cracked the ICAR-JRF examination in 2021. He has received the Best Oral Presentation Award at an International conference. During his professional tenure, he has successfully organized four Capacity Building Programs funded by NFDB, Hyderabad, and MANAGE, Hyderabad.

Prof. S. Felix is a distinguished figure in the field of fisheries education and technology consulting, with a career spanning over four decades. Currently serving as the Dean of the College of Fisheries Science at DIFST (Devasahayam Institute of Fisheries Science and Technology) in Midalam, Kanniyakumari district, Tamil Nadu and also President of AQUAFIN (Aquaculture Technology & Research Foundation India), he has held several prestigious positions, including Vice Chancellor of Tamil Nadu Fisheries University (TNJFU) and President of the World Aquaculture Society (WAS)-APC. Prof. Felix is recognized for his exemplary leadership and contributions to the fisheries and aquaculture sector, demonstrated through his involvement in various committees and organizations. He is a prolific author, having published numerous research papers, books, and manuals, and has played a pivotal role in introducing new Fisheries science courses and establishing Fisheries Colleges and Research Centers.

Fisheries Statistical Methods

Theory and Practical Manual

J. Rujan
S. Felix

College of Fisheries Science,
St. Devasahayam Institute of
Fisheries Science & Technology,
Midalam, Kanniyakumari Dist. 629193 (Tamil Nadu)

2025

Daya Publishing House®
A Division of
Astral International Pvt. Ltd.
New Delhi – 110 002

ISBN: 978-93-5919-230-7

Published by : **Daya Publishing House®**
A Division of
Astral International Pvt. Ltd.
– ISO 9001:2015 Certified Company –
4736/23, Ansari Road, Darya Ganj
New Delhi-110 002
Ph. 011-43549197, 23278134
E-mail: info@astralint.com
Website: www.astralint.com

Preface

In the dynamic world of fisheries, where scientific information dominates decision-making and establishment of policies, a thorough understanding of statistical methods is crucial. This Manual serves as a beacon, demonstrating all the possible paths of statistical analysis in the context of Fisheries Science. From the basic principles of statistics to the advanced uses in estimating marine and inland fish catch, each chapter reveals an important aspect of statistical inference. As providers of aquatic ecosystems face complex challenges that range from sustainable resource management to biodiversity conservation, the tools outlined herein serve as pillars for informed action. Focusing on an extensive range of ideas such as population sampling, probability distributions, and hypothesis testing, this manual provides professionals with the analytical skills required for effective stewardship. With a thorough blend of theory and practical, it spans the gap between statistical theory and real-world fisheries management, allowing readers to confidently navigate the depths of uncertainty. As we begin on an adventure through the statistical seascape of Fisheries Science, let us grasp knowledge's breakthrough power, striving for a future where data-driven insights protect the abundant resources of our oceans and waterways.

J. Rujan
S. Felix

Contents

Chapter-1

Statistics, Population & Data

1. Definition of Statistics and Its Importance in Fisheries Science

Definitions of Statistics:

Statistics is a branch of mathematics that deals with collecting, organizing, analyzing, interpreting, and presenting data. "Statistics" is a commonly employed term derived from the Latin word "status," which originally referred to the collection of knowledge or data.

According to A.L. Bowley, "Statistics is an arithmetical statement of facts in any department of study placed in relative to each other".

Prof. Horace Secrist states, "Statistics is the probable giving to routine morals of accuracy, collected systematically for a per-determined resolution and placed relative to each other."

'Father of Statistics' - Ronald Fisher:

Sir Ronald Aylmer Fisher (1890-1962), renowned as "his time's greatest scientist," was a British statistician and biologist who significantly contributed to experimental design and population genetics. He is regarded as the "Father of Modern Statistics and Experimental Design". He laid the foundation of statistical inference, invented experimental design, randomisation, ANOVA, etc. The first "Statistical Methods for Research Workers" edition appeared in 1925.

Scope of Fisheries Statistics:

The scope of fisheries statistics encompasses the collection, analysis, and reporting of data related to marine and inland fisheries resources, production, and socio-economic aspects. Some key points about the scope of fisheries statistics:

- It covers both marine and inland fisheries, including capture fisheries and aquaculture.
- For marine fisheries, data is collected on catch, fishing effort, stock assessment, economics, and socio-economic aspects of fishing communities.
- For inland fisheries, data is collected on production from different water bodies like rivers, reservoirs, ponds, etc. using appropriate sampling methodologies.
- Fisheries statistics provide information on the exploited fish populations, fishing craft and gear, and fishery-related infrastructure to support management decisions.
- Data is collected at the national level through sample surveys and census, and reported in standard formats to enable comparisons over time and across regions.
- Fisheries statistics underpin the development and utilization of the fisheries sector by providing a reliable database for policy formulation, planning, and management.
- The scope also includes collating information on the socio-economic benefits derived from fisheries and the institutional arrangements for governance.

How are Fisheries Statistics used in Fisheries Management?

(i) **Stock Assessment and Quota Monitoring:** Fisheries statistics provide data on catch, fishing effort, and biological characteristics of fish populations. This information is used in stock assessments to evaluate the health and status of exploited fish stocks. Fishery managers then use this scientific advice to set sustainable catch limits and quotas.

(ii) **Informing Management Decisions:** Fisheries statistics help characterize the commercial fishing fleets, effort, and discards. This data informs the development of fisheries management plans and regulations to ensure the long-term sustainability of fisheries resources.

(iii) **Evaluating Management Measures:** By collecting data over time, fisheries statistics allow managers to evaluate the effectiveness of management measures in achieving conservation and socio-economic objectives. This enables adaptive management to improve outcomes.

(iv) **Supporting Ecosystem-Based Management:** The scope of fisheries statistics is expanding to include data on habitat, bycatch, and interactions with protected species. This supports a more holistic, ecosystem-based approach to fisheries management that considers the interconnectedness of marine systems.

(v) **Enabling International Cooperation:** Standardized fisheries statistics enable comparisons across regions and countries. This facilitates international cooperation in the management of shared and migratory fish stocks, as required by agreements like the UN Fish Stocks Agreement.

Sources of Fisheries Statistics Data in India:

(i) **State Governments and UT Administrations:** Data on fish production, prawn production, fish seed production, disposal of fish catches, and aquaculture are collected from State Governments and UT Administrations in India.

(ii) **Central Marine Fisheries Research Institute (CMFRI):** The Survey for Marine Fisheries methodology developed by CMFRI in Kochi is used for collecting data on marine fish production in India.

(iii) **Central Inland Fisheries Research Institute (CIFRI):** CIFRI in Barrackpore, West Bengal, is involved in developing methodologies for collecting data on inland fisheries in India.

(iv) **Fishery Survey of India (FSI):** Coordinates with other agencies for conducting Fisheries Census in India, covering both inland and marine fisheries.

(v) **National Productivity Council and Other Institutes:** Agencies like the National Productivity Council and other renowned institutes are involved in conducting studies on various aspects of the fisheries sector in India.

(vi) **Marine Export Data:** It is published annually by the Marine Products Export Development Authority, Cochin. It contains information on country-wise exports, region-wise exports, item-wise exports, average unit value, world markets, prices, marine fish landings, etc.

(viii) **Handbook on fisheries statistics:** The Government of India, through its Fisheries Division under the Ministry of Agriculture and Cooperation, periodically releases national fisheries data as an official document for the Central Board of Fisheries meeting.

Sources of Fisheries Statistics Data Worldwide:

- **Central Statistics Office, Ministry of Statistics & Programme Implementation:** Provides data on the contribution of India to world fish production in selected years.
- **Food and Agriculture Organization (FAO):** International organization that compiles and publishes fisheries statistics data from various countries, including India, for comparison and analysis.
- **Secondary Sources:** Various secondary sources like MPEDA-Agri-exchange website are used to collect data for research and analysis on the growth and trade performance of Indian fisheries.

2. Concepts of Population and Sample

1. Population

A population refers to collection of the entire group of individuals, objects, or events that share common characteristics. It includes all elements of a dataset and measurable parameters known as parameters.

Types of Population in Statistics:

- **Finite Population:** A finite population is countable and includes a specific, limited number of elements.

  Example: The total count of fish in a specific pond or the number of fishing boats in a harbor represent finite populations in fisheries statistics.

- **Infinite Population:** An infinite population is uncountable and theoretically limitless.

  Example: The potential number of fish in the ocean or the infinite possibilities of fishing locations represent infinite populations in fisheries statistics.

- **Existent Population:** An existent population consists of concrete, real individuals or entities.

  Example: The actual count of fish species in a particular river or the number of fish farms in operation represent existent populations in fisheries statistics.

- **Hypothetical Population:** A hypothetical population includes units that are not physically present or observable.

  Example: The theoretical estimation of fish populations in a newly discovered underwater ecosystem or the projected growth of a specific fish species represent hypothetical populations in fisheries statistics.

2. Sample:

A sample is a subset of the population, consisting of one or more observations drawn from the population. The measurable characteristic of a sample is called a statistic. The number of individuals or observations in a sample is called the sample size and is generally denoted by 'n'.

There are two broad categories of sampling methods: probability sampling and non-probability sampling.

(a) Probability Sampling:

In probability sampling, each member of the target population has a known chance of being sampled. Examples include:

- **Simple Random Sampling:** Selecting a random sample from the entire population, like randomly selecting fishing sites to estimate fish density in a lake.
- **Stratified Sampling:** Dividing the population into subgroups (strata) and randomly sampling from each stratum. For example, stratifying by geographic region when sampling anglers.
- **Cluster Sampling:** Dividing the population into clusters and randomly selecting some clusters. For instance, grouping fishing access sites by location and randomly selecting some sites to intercept anglers.
- **Systematic Sampling:** Selecting every n^{th} unit from a list. This could be used to sample fish from a catch, like measuring every 10th fish.

(b) Non-Probability Sampling:

In non-probability sampling, the chance of selection is unknown. Examples include:

- **Convenience Sampling**: Selecting readily available subjects, like interviewing anglers at easily accessible fishing sites.
- **Quota Sampling**: Selecting subjects to fill predefined quotas, such as interviewing a certain number of anglers of each fishing mode.
- **Purposive Sampling**: Selecting subjects based on judgment, like choosing fishing sites expected to have high activity.
- **Snowball Sampling**: Subjects recruit future subjects, as in an angler reporting program where each participant recruits another angler.

Sampling Frame:

It is a list, map, or other specification of units that constitute available information regarding the population, forming the basis for drawing samples.

Sampling with Replacement:

Sampling with replacement involves selecting items from a population and returning each item to the population before the next selection. This method allows items to be selected more than once. If a sample of size n has to be drawn from a population of size N, then there will be N^n possible samples in the case of sampling with replacement.

Sampling without Replacement:

In sampling without replacement, the unit once selected at any draw is not replaced back, so that the same unit cannot enter the sample more than once. If a sample of size n has to be drawn from a population of size N, then there will be NC_n possible samples in the case of sampling without replacement.

Characters of samples:

Quantitative characters: Quantitative data are measures expressed as numbers, counts, or values. They deal with numeric variables and answer questions like "how many" or "how much."

- ❑ Examples: Quantitative characteristics in fisheries sampling could include the number of fish caught in a specific area, the weight of fish in a catch, or the length of fish species in a population.

Qualitative characters: Qualitative data are descriptive and expressed in terms of language rather than numerical values. They focus on qualities, traits, or attributes and answer questions like "why" or "how."

- ❑ Examples: Qualitative characteristics in fisheries sampling could involve traits like the species of fish caught, the color patterns of fish, or the behavior of fish in response to environmental changes.

Characteristic	Quantitative	Qualitative
Definition	Numeric, countable, or measurable data	Descriptive, interpretation-based, and language-focused data
Examples	Number of fish caught, weight of fish, length of fish species	Species of fish, color patterns of fish, behavior of fish
Measurement	Numerical values, counts	Descriptive traits, categories
Analysis	Statistical analysis, calculating means, medians, etc.	Grouping data into categories and themes
Research Questions	Answers "how many" or "how much"	Answers "why" or "how"

Characteristic	Quantitative	Qualitative
Approach	Objective, fixed, and universal	Subjective, unique, and interpretation-based
Data Collection Methods	Surveys, experiments	Interviews, focus groups, observation

Constant: A constant is a value that does not change during an experiment or study. It remains fixed and does not vary.

Variable: A variable is a factor that can change or vary in an experiment. It is not fixed and can take on different values. In other words, a quantitative characteristic that varies from individual to individual is called a 'variable'.

- **Continuous variable:** A continuous variable can take on any value within a certain range. It can be measured at any point along a continuum and can have an infinite number of possible values. For example, the length of fish, weight of fish, etc.
- **Discrete variable:** A discrete variable can only take on specific, distinct values. It cannot be measured at every point and has a finite number of possible values. For example, Number of fish discarded, number of fin rays, number of vertebrae, etc.

Attribute: An attribute is a descriptive characteristic or quality of an object or individual that is not quantifiable. Attributes are qualitative in nature and describe features that cannot be measured numerically.

3. Collection of Data

Data collection in statistics refers to the process of gathering, measuring, and analyzing accurate data to respond to specific research questions, test hypotheses, and assess results. It involves collecting information from relevant sources to resolve the study topic and evaluate the outcome.

There are two main types of data collection methods:

1. Primary Data Collection:

- Primary data is collected from first-hand experience and is not used in the past.
- It is highly accurate and specific to the research's motive.
- Primary data collection methods can be further divided into quantitative and qualitative methods.

2. Secondary Data Collection:

- Secondary data is information gathered from secondary sources, providing second-hand information in the form of commentaries, reviews, and comparisons.
- It indicates that someone has previously analyzed and reviewed the material and made the data of their study available.
- Sources of secondary data include books, journals, periodicals, newspapers, and online databases.

1. Primary Data Collection Methods:

(i) Census or complete enumeration

(ii) Sample survey

(i) Census or Complete Enumeration

Collecting data from every unit or element of the population. When studying specific characteristics of a population, it is advisable to measure all individuals, a scenario where census enumeration is typically adopted.

Advantages:

- **Universal Coverage:** A census aims to cover the entire population, providing a comprehensive understanding of the population's characteristics.
- **Intensive Study:** It provides in-depth information on various facets of the population, including demographics, housing, social, and economic characteristics.
- **Accurate and Reliable Results:** Since every unit of the population is included, the conclusions drawn from a census are more accurate and reliable.
- **Benchmark Data:** Census data serves as a benchmark for subsequent surveys and studies, allowing for comparisons and tracking changes over time.
- **Useful for Small Areas:** Census data can provide information for small areas, which is often not possible with sample surveys.
- **No Sampling Error:** As the entire population is surveyed, there is no sampling error, which can be a significant advantage in certain studies.

Disadvantages:

- **Costly:** Conducting a census is usually more expensive due to its comprehensive nature, requiring significant resources and manpower.
- **Time-Consuming:** A census can take years to complete, especially for national censuses, which can delay the availability of results.
- **Non-Sampling Errors:** Although a census is generally more accurate, it is also prone to non-sampling errors, such as inaccuracies in data collection.
- **Not Suitable for Large or Infinite Populations:** A census is impractical for large or infinite populations, where the number of items is unlimited.
- **Difficult to Achieve 100% Response:** It can be challenging to achieve a 100% response rate, which can lead to non-response errors.
- **Data Analysis Challenges:** Analyzing and interpreting the large volume of data collected during a census can be a complex and time-consuming task.

(ii) Sample Survey Method

Collecting data from a representative subset (sample) of the population. When dealing with large populations, this method is commonly employed for investigations, with results generalized to the entire population.

Sample survey is the only logical alternative to census enumeration in the following situations:

- While dealing with infinite populations. For example, to study the characteristics of a fish population in a sea, it is not possible to collect and measure all the fish from sea and have to be satisfied with a sample.
- If study requires destroying of units of individuals of the population. For instance, in estimation of 'fecundity', finding 'biochemical composition' of fish etc., sampling is the only alternative.

Advantages:

- **Lower Cost:** Conducting a sample survey is significantly less expensive compared to a census, as it involves collecting data from a smaller subset of the population.
- **Faster Data Collection:** Sample surveys can be completed in a shorter time frame than a census, as fewer units need to be surveyed.
- **Flexibility:** Sample surveys allow for more flexibility in terms of the types of questions that can be asked and the methods of data collection that can be used.
- **Ability to Study Complex Phenomena:** Sample surveys can be used to study complex phenomena in depth, as they allow for the collection of more detailed and nuanced data.
- **Ability to Generalize:** If the sample is representative of the population, the results of a sample survey can be generalized to the larger population with a known degree of accuracy.
- **Ability to Conduct Pilot Studies:** Sample surveys can be used to conduct pilot studies to test the feasibility and effectiveness of a larger-scale study before investing significant resources.

Disadvantages:

- **Sampling Error:** Sample surveys are subject to sampling error, which is the difference between the sample statistic and the true population parameter.
- **Non-Sampling Error:** Sample surveys are also subject to non-sampling errors, such as coverage error, measurement error, and non-response error, which can be difficult to quantify and control.
- **Difficulty in Selecting a Representative Sample:** Selecting a truly representative sample can be challenging, especially when the population is heterogeneous or when the units of the population are not easily accessible.
- **Potential for Bias:** Sample surveys can be subject to various types of bias, such as selection bias, response bias, and interviewer bias, which can affect the validity and reliability of the results.

- Difficulty in Replicating: It can be difficult to replicate the results of a sample survey, as the sample may not be easily accessible or may have changed over time.

(iii) Difference between Census and Sample Survey:

Aspect	Census	Sample Survey
Scope	All the items related to problems are analyzed.	Only a small group of items is chosen from a large group.
Cost	It requires more money, time, and energy.	It requires less money, time, and energy. It is more economical.
Accuracy and reliability	Results will be more accurate and reliable as all the units of the universe are studied.	Conclusions drawn are less accurate and reliable.
Organizations	It needs more organizational skill and a large force of enumerators.	It needs less organizational skill and a fewer number of enumerators.
Deep study	Deep study of the subject is possible.	Deep study of the subject is not possible.
Suitability	The method is suitable when more accuracy is required and the area of enquiry is limited.	This method is suitable when the area of enquiry is wide and extensive study is not required.

2. Secondary Data Collection

Secondary data collection refers to gathering information that has already been collected by someone else for a different purpose. The sources of secondary data collection can be broadly classified into two categories:

(i) Published Sources

1. Books, journals, magazines, and newspapers
2. Government publications like census data, economic surveys, and statistical abstracts
3. Reports and publications of trade associations, chambers of commerce, and international organizations
4. University research publications
5. Online databases and websites

(ii) Unpublished Sources

1. Data and records maintained by companies, both public and private
2. Unpublished dissertations and theses
3. Personal records and correspondence
4. Data collected by market research firms and consulting organizations

Advantages of using secondary data:

- Readily available and less time-consuming to collect
- Often less expensive compared to primary data collection
- Provides historical trends and longitudinal data
- Allows for comparisons across time periods and geographic regions

Disadvantages of secondary data:

- Data may not be a perfect fit for the current research question
- Lack of control over data quality and collection methods
- Difficult to assess reliability and validity of data collected by others
- Potential for data being outdated or obsolete

3. Data in Statistics

Data in statistics refers to the individual pieces of factual information recorded and used for analysis.

Types of Data:

Quantitative Data in Fisheries

- **Discrete Data**: Represents counts of individual items or values that are whole numbers, such as the number of fish caught in a net.
- **Continuous Data**: Involves measurements of continuous or non-finite values, like the length of fish species.

Qualitative Data / Categorical Data in Fisheries

(i) **Nominal Data:** It is qualitative data that categorizes variables into distinct, non-ordered groups without numerical value. It is used for labeling variables, such as fish species or fishing gear types, without implying any quantitative relationship.

(i) **Ordinal Data**: It is a type of data where values follow a natural order or rank, but the differences between values cannot be determined or are meaningless.

4. Presentation of Data

Presentation of Data

Once data are collected, the first objective should be to summarize and present them in a form that highlights their main characteristics. Tables, diagrams, and graphs are important forms of presenting statistical data.

Classification of data

Classification is the process of arranging data into sequences according to their common characteristics or separating them into different related parts (Professor. Secrist). If a marine biologist aims to assess the ecological well-being of a specific region's fish population, they can achieve this by collecting data on the various fish species in that area and then categorizing them based on factors like species diversity, age distribution, habitat preferences, and reproductive rates.

Objectives of Data Classification

- ❑ Simplification and Organization
- ❑ Summarization of large data sets
- ❑ Facilitating Interpretation
- ❑ Comparisons and Contrasts
- ❑ Prediction and Inference

Methods of Classification

Geographical Classification	When data are classified with reference to geographical locations such as countries, states, cities, districts, etc., it is known as geographical classification. It is also known as 'spatial classification'.
Chronological Classification	A classification where data are grouped according to time is known as a chronological classification. In such classification, data are classified either in ascending or in descending order with reference to time such as years, quarters, months, weeks, etc. It is also known as temporal classification'.
Qualitative Classification	Under this classification, fish data are categorized based on specific attributes or characteristics such as species type, size, sex, habitat preference, and reproductive behavior. For example, fish may be classified as male or female; fishing units may be classified as trawlers, gill netters, long liners etc.
Quantitative Classification	This type of classification is made on the basis of some measurable characteristics like height, weight, age of fishes, income of fishers, etc. 'An important form of quantitative classification is 'frequency distribution' table.

Frequency Distribution Table

- In statistics, a frequency distribution table is a comprehensive way of representing the organization of raw data of a quantitative variable. This table shows how various values of a variable are distributed and their corresponding frequencies.
- It consists of dividing the range of observations into intervals (usually of equal size) and noting the number of observations falling into each interval (called frequency) and presenting it in a tabular form.
- The data presented in the form of a frequency distribution table are called 'grouped data'.

Terminologies

Class interval and Class Limits

A symbol defining a class 10-15 is called a class interval. The end numbers 10 and 15 are called class limits; the smaller number 10 is called the lower class limit, and the larger number 15 is called the upper class limit. The terms 'class' and 'class interval' are often used interchangeably.

Inclusive and Exclusive Class Intervals

The class interval for instance 10-15 will be called inclusive class interval if it includes all the values from 10 to 15 including the lower limit 10 and the upper limit 15. On the other hand, it will be called exclusive class interval if it does not include the upper limit 15.

The Size or Width of Class Interval

The class width is defined as the difference between the lower and upper class boundaries. In the case of equal class intervals, it represents the difference between two successive lower class limits or two successive upper class limits.

Class Boundaries or True Class Limit

In inclusive class intervals, the class boundaries are obtained by adding the upper limit of one class interval to the lower limit of the next higher class interval and then dividing by 2. In the case of exclusive class intervals, class limits coincide with class boundaries.

For Example

In the inclusive form: 4–8, 9–13, 14–18

True lower limit of the class = lower limit -0.5

True upper limit of the class = upper limit -0.5

Hence, the true class limit for 4-8 are 4.5 are & 8.5.

In the exclusive form: 10–20, 20–30, 30–40

In the exclusive form, the upper and lower limits of a class are respectively known as the true upper limit and the true lower limit.

Class Mark or Class Mid-point:

The class mark is the midpoint of the class interval and is obtained by adding the lower and upper class limits and then dividing by 2. For example, the class mark of the interval 10 to 15 is calculated as (10 + 15) / 2 = 12.5. The class mark is also referred to as the class midpoint.

Steps in the formation of frequency distribution table:

- ❑ Determine the largest and the smallest numbers data and then find the range by subtracting the given smallest number from the largest number.
- ❑ Divide the range into a convenient number of class intervals having the same size.
- ❑ Determine the number of observations' falling into each class interval, i.e., find the frequency. This is done by using tally marks.

General rules to be followed while forming a frequency distribution table:

- ❑ Class intervals should be clearly defined.
- ❑ Avoid using open-ended class intervals.
- ❑ Whenever possible, use class intervals of equal size (width) to facilitate computations.
- ❑ It is convenient to choose the midpoint of a class interval as a whole number, i.e., an integer.

Example

The lengths (in cm.) of 30 randomly selected fish are given below.

5	26	18	21	36	37	39	26	31	33
22	26	15	12	20	21	48	27	25	30
28	35	38	36	41	25	26	35	20	21

Arrange the observations into a frequency distribution table and also find out the relative frequency distribution.

Solution

Step 1: Determine the range of the data.

- ☆ Calculate the range by subtracting the minimum value from the maximum value. Range = Maximum value - Minimum value = 48 - 12 = 36 cm

Step 2: Decide on the number of intervals (bins) and their width.

- ☆ Choose the number of intervals. Let's select 6 intervals for this dataset.
- ☆ Calculate the interval width by dividing the range by the number of intervals.
- ☆ Interval width = Range / Number of intervals = 36 / 6 = 6 cm

Step 3: Determine the boundaries for each interval.

- ☆ Start with the lower limit of the first interval, which is the minimum value.
- ☆ Determine the upper limit of the last interval, which is the maximum value.
- ☆ Divide the range into equal intervals based on the interval width.
- ☆ **First interval:** 12 - 17 cm
- ☆ **Second interval:** 18 - 23 cm
- ☆ **Third interval:** 24 - 29 cm
- ☆ **Fourth interval:** 30 - 35 cm
- ☆ **Fifth interval:** 36 - 41 cm
- ☆ **Sixth interval:** 42 - 48 cm

Step 4: Count the frequency of observations falling into each interval.

Class Interval	Frequency
12 - 17	2
18 - 23	5
24 - 29	6
30 - 35	8
36 - 41	5
42 - 48	4

Step 5: Calculate the relative frequency distribution.

Relative frequency = Frequency / Total number of observations

= Frequency / 30

Create a table with columns for the class intervals, frequencies, and relative frequencies. This gives the frequency distribution table

Class Interval	Frequency	Relative Frequency
12 - 17	2	2/30 ≈ 0.067
18 - 23	5	5/30 = 0.167
24 - 29	6	6/30 = 0.200
30 - 35	8	8/30 ≈ 0.267
36 - 41	5	5/30 = 0.167
42 - 48	4	4/30 ≈ 0.133

Relative Frequency Distribution

If the relative frequencies of each class are expressed as percentages by multiplying by 100, the resulting distribution is called a percentage frequency distribution. The relative frequency distribution or percentage frequency distribution is especially useful for comparing two or more sets of data that have different numbers of observations (total frequency).

Cumulative Frequency Table

- The cumulative frequency is calculated by adding each frequency from a frequency distribution table to the sum of its preceding frequencies.
- The last value will always be equal to the total number of observations.

Types

Less Than Cumulative Frequency

The less than cumulative frequency distribution is obtained by successively adding the frequencies of all the previous classes up to and including the class against which it is written. In this type, the cumulative frequency starts from the lowest size and progresses to the highest size.

Greater Than Cumulative Frequency

The greater than cumulative frequency is also known as the more than type cumulative frequency. In this type, the cumulative frequency distribution is obtained by determining the cumulative total frequencies starting from the highest class down to the lowest class.

Diagrammatic and Graphical Representation of Data

Diagrams and graphs serve as visual representations of mathematical concepts and data. They are easily understood by the general population, and the impressions they create tend to be long-lasting. Additionally, they facilitate the comparison of trends, values, and relationships. Graphs are particularly useful for locating statistical measures such as the median, quartiles, and other partition values. However, despite these advantages, it's important to note that diagrams and graphs are simply visual aids and can only supplement, not replace, tables or the original data.

Advantages of Diagrammatic Presentation

- Makes it much easier to understand
- Presentation is much simpler
- Comparison operations are more interactive
- Accepted universally
- Improves the representation of data as a whole

Types of Diagrams

- Bar Diagrams
- Pie-Diagrams
- Histogram
- Frequency polygon
- Frequency curve
- Ogives

Bar Diagram

- Many sources consider William Playfair (1759-1824) to have invented the bar chart. The graph depicting the Exports and Imports of Scotland to and from different parts for one Year from Christmas 1780 to Christmas 1781, found in his work 'The Commercial and Political Atlas,' is considered to be the first bar chart in history.
- A bar diagram is a diagram that presents categorical data using rectangular bars, with heights or lengths proportional to the values they represent.
- A bar graph visually represents comparisons among discrete categories using rectangular bars, with the length or height of each bar proportional to the values they represent.

Properties of Bar Diagram

- The Bars can be used in vertical and horizontal directions. A vertical bar chart is sometimes called a line column graph.
- These Bars all have a uniform width.
- All the Bars have a common base.

Simple Bar Diagram

If the classification is based on attributes and if the attributes are to be compared with respect to a single character, we use a simple bar diagram. Example,

- The fish production of different years.
- The yield performance of different fish species.
- Effect of different feed formulations, etc.

Simple bar diagrams consist of vertical bars of equal width. The heights of these bars are proportional to the volume or magnitude of the attribute. All bars stand on the same baseline. The bars are separated from each other by equal intervals. The bars may be colored or marked.

Example: The following data represent the total fish production in India from 2016-2022. Represent the data by bar diagram.

Year	Production (Lakh Tonnes)
2016-17	114.31
2017-18	127.04
2018-19	135.73
2019-20	141.64
2020-21	147.25
2021-22	162.48

Solution

To represent the total fish production in India from 2016-2022 using a bar diagram, you would follow these steps:

- Create a bar graph with the years on the x-axis and the production (in lakh tonnes) on the y-axis.
- Each bar represents a year, and its height corresponds to the production value for that year.
- Use the provided data to plot the bars for each year.

Here is how the bar diagram for the given data represented,

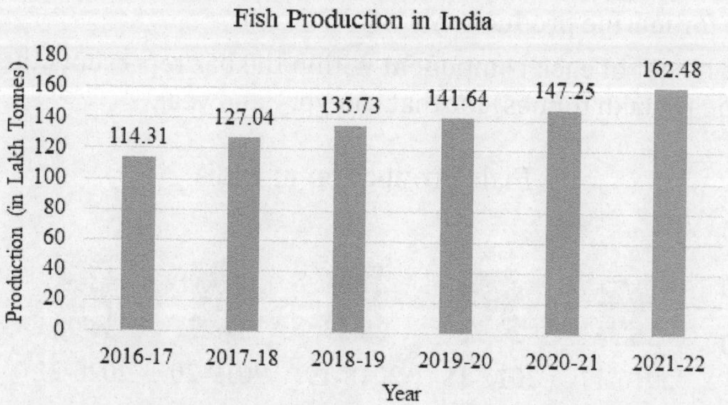

Component or Subdivided

- If the variable or character under study has two or more component parts, then the component bar diagram is used.

- ❑ This is also called sub – divided bar diagram. Instead of placing the bars for each component side by side we may place these one on top of the other.
- ❑ These bars are subdivided into component parts and are marked in different colors to distinguish the components from one another.
- ❑ This diagram is more useful when one wants to compare the size and also the relation between each component and the total.

Example

The following data represent the total fish production in India from 2016-2022. Draw component bar diagram.

Year	Inland (lack Tonnes)	Marine (Lack Tonnes)
2016-17	36.25	78.06
2017-18	37.56	89.48
2018-19	38.53	97.20
2019-20	37.27	104.37
2020-21	34.76	112.49
2021-22	41.27	121.21

Solution

To draw a component bar diagram for the total fish production in India from 2016-2022, the data is represented by two sets of bars for each year, one for inland production and one for marine production.

- ❑ Draw a horizontal axis representing the years (2016-17, 2017-18, ..., 2021-22).
- ❑ Draw a vertical axis representing the total fish production (in lakh tonnes).
- ❑ Each bar is divided into two components: one for inland production and one for marine production.
- ❑ The height of each component within the bar represents the production value (in lakh tonnes) for that category and year.

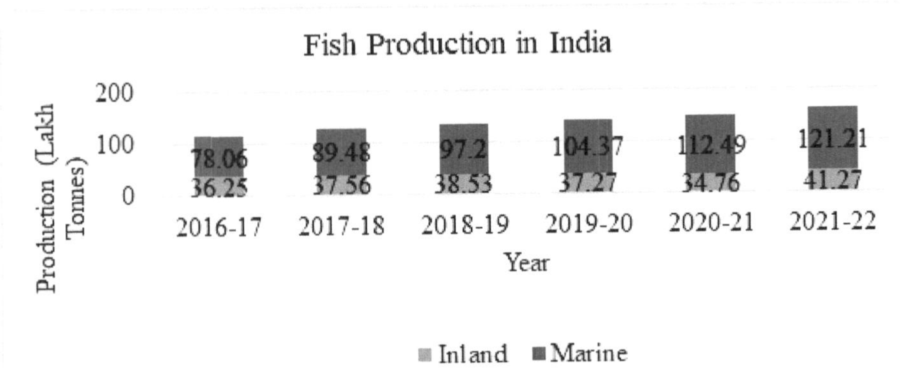

Multiple Bar Diagram

- When there are two or more different comparable sets of closely related variables or characters, multiple bar diagram is used.
- If only two characters are to be compared within each attribute, then the resultant bar diagram used is known as double bar diagram.
- The diagram consists of multiple bars drawn continuous to one another representing the component parts and then marked in different colors to distinguish the components.

Example

The following data represent the total fish production in India from 2016-2022. Draw multiple bar diagram.

Year	Inland (Lack Tonnes)	Marine (Lack Tonnes)
2016-17	36.25	78.06
2017-18	37.56	89.48
2018-19	38.53	97.20
2019-20	37.27	104.37
2020-21	34.76	112.49
2021-22	41.27	121.21

Solution

To draw a multiple bar diagram for the total fish production in India from 2016-2022, with separate bars for inland and marine production, follow these steps:

- Create a horizontal axis representing the years (2016-17, 2017-18, ..., 2021-22).
- Create a vertical axis representing the total fish production (in lakh tonnes).
- Draw two sets of bars for each year: one for inland production and one for marine production.
- The height of each bar represents the production value for that year and category.

Here's the multiple bar diagram based on the provided data:

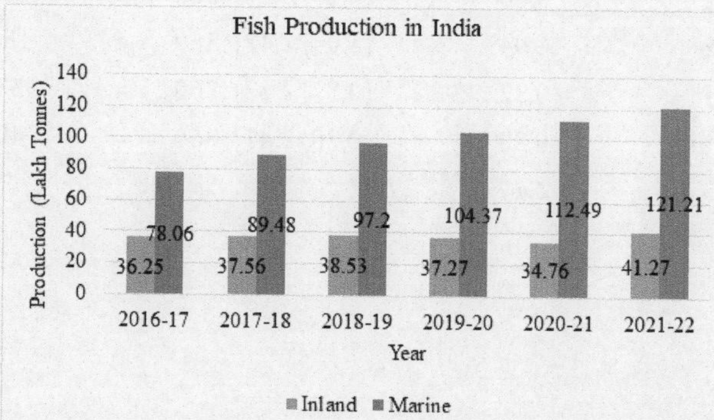

Pie Diagram

- William Playfair's Statistical Breviary of 1801 includes two such graphs. Playfair presented an illustration that contained a series of pie charts.
- A pie chart (or circle chart) is a circular statistical graphic divided into slices to illustrate numerical proportion.
- In a pie chart, the arc length of each slice (and consequently its central angle and area) is proportional to the quantity it represents.
- Pie charts are very widely used in the business world and the mass media.

Construction of Pie Diagrams

Step 1: Whenever one set of actual value or percentage are given, find the corresponding angles in degrees using the following formula

$$Angle = \frac{\text{Actual Value} \times 360}{\text{Total Value}}$$

Step 2: Find the radius using the area of the circle πr2 where value of π is 22/7 or 3.14

Example:

The following data represent the species wise fish production in India in 2022. Draw a pie diagram.

Species	Sharks	Skates	Rays	Oil Sardine	Hilsa Shad	Stolephorus
Production (Lakh Tonnes)	0.27	0.33	0.18	1.04	0.37	0.28

Solution

To construct a pie diagram, the fact that the total 2.47 LT corresponds to 360^0 is used. Thus the angle of degree required for different species are

Species	Production				Angle
Sharks	0.27	:	(0.27/2.47) 360	=	39.35
Skates	0.33	:	(0.33/2.47) 360	=	48.10
Rays	0.18	:	(0.18/2.47) 360	=	26.23
Oil Sardine	1.04	:	(1.04/2.47) 360	=	151.58
Hilsa Shad	0.37	:	(0.37/2.47) 360	=	53.93
Stolephorus	0.28	:	(0.28/2.47) 360	=	40.81
Total	**2.47**				360

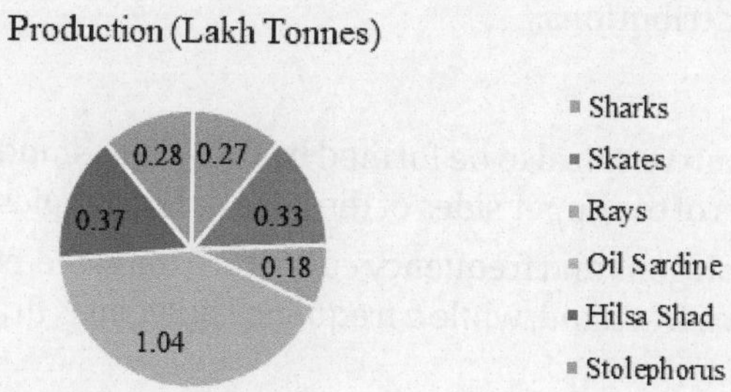

Histogram

- A histogram is a graphical representation of a grouped frequency distribution with continuous classes.
- It is an area diagram and can be defined as a set of rectangles with bases along with the intervals and with areas proportional to frequencies in the corresponding classes.
- As the adjacent interval leave no gaps, the rectangles of a histogram touch each other to indicate that the original variable is continuous.
- If the frequency distribution has inclusive class intervals, first convert into exclusive type and draw the histogram.
- Histograms are sometimes confused with bar charts. A histogram is used for continuous data, where the bins represent ranges of data, while a bar chart is a plot of categorical variables. Some authors recommend that bar charts have gaps between the rectangles to clarify the distinction.

Frequency Polygon

- A frequency polygon can be obtained by joining the midpoints of the upper sides of the adjacent rectangles in the histogram with small straight lines.
- A frequency polygon is sometimes left open at each end down to the horizontal axis (X-axis), and the points chosen for closing the polygon are the midpoints of the first class interval on either end of the distribution with zero frequency.
- If the class intervals are of unequal width, it is obtained by plotting the frequency density against the class midpoints.

Uses

- Frequency polygons are a graphical device for understanding the shapes of distributions.
- They serve the same purpose as histograms, but are especially helpful for comparing sets of data.
- Frequency polygons are also a good choice for displaying cumulative

frequency distributions.

Frequency curve:

- A frequency curve can also be formed by drawing a smooth curve through the midpoints of the upper sides of the adjacent rectangles of the histogram.
- Frequency polygons and frequency curves are the same, except a frequency curve is drawn freehand, while a frequency polygon is drawn using a scale.

Ogive curves:

The graph of cumulative frequency distribution is known as ogive. Class boundaries are taken on X-axis. At each class boundary, the corresponding cumulative frequency is marked, and these points are joined by a smooth curve.

There are two types of Ogives:

- **Less than type ogive:** Plotting the less than cumulative frequencies against the upper class boundaries.
- **More than type ogive:** Plotting the more than cumulative frequencies against the lower class boundaries.

Merits of Diagrams and Graphs:

- ☆ Diagrams attract most people.
- ☆ Technical knowledge or education is not necessary.
- ☆ Less time and effort are required.
- ☆ Diagrams present data in a proper perspective.
- ☆ Diagrams leave a lasting impression.
- ☆ Language is not a barrier.
- ☆ A widely used tool.

Demerits of Diagrams and Graphs

- ☆ Diagrams are approximations.
- ☆ Minute differences in values cannot be accurately represented in diagrams.
- ☆ Large differences in values can distort the appearance of the diagram.
- ☆ Some diagrams, such as pie charts, require expertise to create effectively.
- ☆ Different scales can convey different interpretations to laymen.

Chapter-2

Central Tendency, Dispersion & Curve

5. Measures of Central Tendency

In biological characteristics, generally, the observations taken on a group of individuals will not be equal, but show a general tendency to cluster around a particular value. This value around which the observations tend to cluster is called that group's 'central tendency' or 'central position'. The central tendency is the statistical measure representing the single value of the entire distribution or a dataset. It aims to accurately describe all the data in the distribution. The central tendency of the dataset can be found using the three important measures.

1. Arithmetic Mean
2. Median
3. Mode

1. Arithmetic Mean:

- The mean represents the average value of the dataset.
- It can be calculated as the sum of all the values in the dataset divided by the number of values.

(i) Calculation for ungrouped data:

In an ungrouped data the arithmetic mean is simply the total of all the values divided by the number of observations.

Thus if there are 'n' observations, x1, x2,...., xn, then the arithmetic mean usually denoted by , $\overline{x}$ will be

$$\overline{x} = \frac{X_1 + X_2 + \ldots + X_n}{n}$$

$$= \frac{\sum X}{n}$$

Example:

Calculate the mean fish length of following

20, 18, 22, 16, 19, 21, 17, 20, 18, 23

Solution

Total no. of. observations, n=10

$$Mean\ \bar{x} = \frac{X_1 + X_2 + \ldots + X_n}{n}$$

$$\bar{x} = \frac{20+18+22+16+19+21+17+20+18+23}{10}$$

$$\bar{x} = 19.4$$

(ii) Calculation for grouped data:

Suppose that the data are given in the form of a frequency distribution in k classes. Let $x_1, x_2, \ldots, x_k$ represent the mid points of 1st, 2nd, ..., kth class, $f_1, f_2, \ldots, f_k$ be their frequencies respectively. Then the arithmetic mean is calculated by the formula.

$$\bar{x} = \frac{f_1 x_1 + f_2 x_2 + \ldots + f_k x_k}{f_1 + f_2 + \ldots + f_k}$$

$$\bar{x} = \frac{\sum fx}{n}$$

Where n = $\sum f$

Example

Find the mean of the following data.

Class Interval	0 - 10	10 - 20	20 - 30	30 - 40	40 - 50
Frequency (fi)	9	13	8	15	10

Solution

The first step is to create the table with the midpoint or marks and the product of the frequency and midpoint. To calculate the midpoint we find the average between the class intervals by using the formula mentioned above.

Midpoint x_i = 0 - 10 = 5 $\left(\frac{10+0}{2}\right)$, 10 - 20 = 15 $\left(\frac{20+10}{2}\right)$ and so on.

$x_i f_i$ = For the class interval 0 - 10 = 5 × 9 = 45, For the class interval 10 - 20 = 13 × 15 = 195 and so on.

Interval	Frequency (f_i)	Class Mark (x_i)	$x_i f_i$
0 - 10	9	5	45
10 - 20	13	15	195

Interval	Frequency (f_i)	Class Mark (x_i)	$x_i f_i$
20 - 30	8	25	200
30 - 40	15	35	525
40 - 50	10	45	450
Total	55		1415

Once we have determined the totals, let us use the formula to calculate the estimated mean.

Estimated Mean $\bar{x} = \frac{\sum f_i x_i}{\sum f_i} = \frac{1415}{55} = 25.73$

Short Method:

The process of computation of the arithmetic mean from the frequency table can be shortened by changing the midpoints of the class intervals to a coded value when the class intervals are of equal size.

This is done by assigning the value zero to any mid-point (class mark), preferably one corresponding to the middle of the frequency distribution.

The midpoints of successive class intervals above and below this midpoint are coded as minus or plus deviations, denoted by 'd' of 1, 2, 3, or more intervals.

The mean is calculated by summing the products obtained by multiplying the frequency in each class by the corresponding d value of the class interval and dividing their sum by 'n' the number of observations.

Since this average is in coded values, it needs to be converted to the original units by multiplying it by the width of the class interval 'h' and then added to the midpoint of the interval selected as zero.

If 'A' is used to represent the mid-point to which the value of zero was assigned, the formula for the arithmetic mean, calculated by this method is,

$$\bar{x} = A + h\frac{\sum fd}{n}$$

Example

Find the mean of the following data.

Class Interval	50 - 70	70 - 90	90 - 110	110 - 130	130 - 150	150 - 170
Frequency	15	10	20	22	16	17

Solution

The first step is to create the table with the midpoint or marks and the product of the frequency and midpoint. To calculate the midpoint we find the average between the class intervals.

Midpoint $x_i = 50 - 70 = 60 \left(\frac{70+50}{2}\right)$, $70 - 90 = 80 \left(\frac{90+70}{2}\right)$, and so on. From the midpoint let us select the assumed mean, so A = 100 and the value of h = 20 which is the class size.

Find the value, where d_i = Let us find the value for each class interval.

Class Interval	f_i	x_i	$d_i = x_i - A / h$	$f_i d_i$
50 - 70	15	60	- 2	- 30
70 - 90	10	80	- 1	- 10
90 - 110	20	100	0	0
110 - 130	22	120	1	22
130 - 150	16	140	2	32
150 - 170	17	160	3	51
Total	**100**			65

Once we have determined the totals, let us use the assume mean formula to calculate the estimated mean.

$$\text{Mean} = x = A + h\frac{\sum f_i d_i}{\sum f_i}$$

$$100 + 20\left(\frac{65}{100}\right)$$

$$= 100 + 13$$

$$= 113.$$

Therefore, the mean of the data is 113.

Merits	Demerits
• It is rigidly defined • It is easy to understand and easy to calculate • It is based upon all values of the given data • It is capable of further mathematical treatment	• It cannot be calculated if any observations are missing. • It cannot be calculated for the data with open end classes. • It is affected by extreme values.

2. Median

- ❑ The median is defined as the middle value when the observations are arranged in ascending or descending order of magnitude.
- ❑ The median divides the data into two equal halves, with half the number of observations lying below it and half lying above it.

(i) Calculation from Ungrounded Data

To find out the median from ungrouped data, it is first necessary to arrange the values in the ascending or descending order of magnitude, with serial numbers 1,2,...n, where n, stands for the total number of observations in the given data.

- ❑ If 'n' is an odd number,

$$Median = \frac{n+1}{2}$$

- If 'n' is an even number,

$$Median = \frac{n}{2} and \frac{n}{2}+1$$

Example:

The data given represents the length of different fishes. Identify the median for the same: 15, 26, 18, 21, 36, 37, 39, 26, 31, 33, 22, 26, 15, 12, 20, 21, 48, 27, 25, 30, 28, 35, 38, 36, 41, 25, 26, 35, 20, 21.

Solution

Arrange the fish lengths in ascending order: 12, 15, 15, 18, 20, 20, 21, 21, 21, 22, 25, 25, 26, 26, 26, 26, 26, 27, 28, 30, 31, 33, 35, 35, 36, 36, 37, 38, 39, 41, 48

- Since there are 31 observations, the median is the 16th value.
- The 16th value in the sorted dataset is 26.
- Therefore, the median fish length is 26 cm.

(ii) Calculation for Grouped Data

Step 1. Make a table with 3 columns. First column for the class interval, the second column for frequency (f), and the third column for cumulative frequency (cf).

Step 2. Write the cumulative frequency in column cf.

Step 3. Find the sum of frequencies, ∑f. It will be the same as the last number in the cumulative frequency column.

Step 4. Find n/2. Then find the class whose cumulative frequency is greater than and nearest to n/2. This is the median class.

Step 5. Now we use the formula

$$Median = L_1 + \frac{\left(\frac{n}{2} - f_c\right)h}{f}$$

Where,

L_1 = Lower class boundary of the median class

n = Total Frequency

f_c = Cumulative frequency before entering the median class

f = The frequency of the median class

h = Width of the class interval

Example:

The marks obtained in Aquaculture exam by 17 students were recorded. What is the median marks of the students?

Marks out of 50	Number of students (f)
0-10	2
10-20	4
20-30	5
30-40	4
40-50	2

Solution

Find the cumulative frequency value for the given data

Marks out of 50	Number of students (f)	Cumulative frequency
0-10	2	2
10-20	4	6
20 30	5	11
30-40	4	15
40-50	2	17

In the grouped data given above, it is not possible to locate the middle-most observation by simple observation of the cumulative frequency column. The 9th observation is the middle-most and its value is somewhere inside the class interval 20-30. To come at this conclusion, we find the value $\frac{n}{2}$ and then locate the class whose cumulative frequency is greater than the value of $\frac{n}{2}$. In the table = 17/2=8.5. Here, 20-30 is the class whose cumulative frequency is 11 which is greater than 8.5.

Median Class

The class whose cumulative frequency than the value is called the median class.

So in the table, the median class is 20-30. To find the median of a grouped data, we have the formula

$$Median = L_1 + \frac{\left(\frac{N}{2} - f_c\right)h}{f}$$

From the table above, we have L_1=20, N=17, fc=6, f=5 and h=10

$$\text{Median} = 20 + \frac{\left(\frac{17}{2}-6\right)10}{5}$$

$$\text{Median} = 20 + \frac{25}{5} = 20+5 = 25$$

Therefore, the median mark of the students is 25.

Merits	Demerits
• It is rigidly defined. • It is easy to understand and easy to calculate. • It is not affected by extreme values. • It can be located just by inspection in many cases. • It can be located graphically. • It can be calculated for data based on ordinal scale.	• It is not based upon all values of given data. • For larger data size the arrangement of data in the increasing order is difficult process. • It is not capable of further mathematical treatment. • It is insensitive to some changes in the data values.

Important Points

- There is only one median for given set of data.
- The median is not affected by extreme values.
- It takes into account only the number of observations, not their size.
- Hence, the median is less reliable than the arithmetic mean and thus, it is less commonly used.

3. Mode

- The mode is the value that occurs most frequently in a dataset.
- The mode may not be unique; there can be multiple modes in a dataset. A distribution with two modes is called bimodal, and a distribution with more than two modes is called multimodal.

(i) Calculation for Ungrouped Data

When the data are not grouped, the mode can be located by arranging the observations in either ascending or descending order. This arrangement helps identify the value that repeats the most.

Example

Weight of fish sampled from a pond are given below. Find their mode

10,15,20,25,15,20,20,25,30,35,20,25,25,30,35,35,35

Solution

To find the mode for this ungrouped data:

- Identify the value(s) that appear most frequently in the dataset.
- In this case, the value 20 appears most frequently (5 times), making it the mode.

So, the mode for the fish weight ungrouped data is 20.

(ii) Calculation for Grouped Data (Continuous Frequency Distributions)

- The class interval which has the maximum frequency is known as the modal class. The modal value lies in this class interval.

- This following formula is used for calculating the mode of the frequency distribution.

$$Mode = L_1 + \frac{(f_m - f_1)h}{(2f_m - f_1 - f_2)}$$

Where,

L_1 = Lower class boundary of the modal class

f_1 = Frequency of the class previous to the modal class

f_2 = Frequency of the class next to the modal class

f_m = Frequency of the modal class

h = Width of the interval

- If the mean and the median have already been calculated, then the following empirical relationship can be used to calculate the mode of the moderately asymmetrical distribution.

Mode = 3 Median – 2 Mean

Example

The weights (in gm), of 50 fishes are recorded as shown below. Calculate their mode

Weight (in gm)	125-130	130-135	135-140	140-145	145-150
Number of fishes	7	14	10	10	9

Solution

Here, the maximum frequency is 14 and the corresponding class is 130-135. So, 130-135 is the modal class such that L_1=130, h=5, f_m=14, f_1=7 and f_2=10.

$$Mode = L_1 + \frac{(f_m - f_1)h}{(2f_m - f_1 - f_2)}$$

$$= 130 + \frac{(14-7)5}{(2\times14-7-10)}$$

$$= 130 + \frac{7\times5}{11}$$

$$= 130 + 3.18$$

$$Mode = 133.18$$

(iii) Calculation for Grouped Data (Discrete Frequency Distributions)

- The frequencies in column (i) represent the original frequencies.
- Column (ii) is created by combining the frequencies two by two.
- To obtain column (iii), we combine the remaining frequencies two by two, leaving out the first frequency.
- Column (iv) is formed by combining the frequencies three by three.

- ❑ Column (v) results from combining the frequencies three by three after omitting the first frequency.
- ❑ Column (vi) is derived by combining the frequencies three by three after leaving out the first two frequencies.
- ❑ List the maximum frequency in each column and determine the corresponding values of 'x'.
- ❑ Identify the 'value of x' that occurs the most frequently.

Merits	Demerits
• It is easy to understand and easy to calculate • It is not affected by extreme values or sampling fluctuation • Even if extreme values are not known mode can be calculated • It can be located just by inspection in many cases • It is always present within the data • It can be located graphically • It is applicable for both qualitative and quantitative data	• It is not rigidly defined • It is not based upon all values of the given data • It is not capable of further mathematical treatment

Important Points

- ❑ The mode is particularly useful in the study of typical size.
- ❑ It is not affected by extreme values.
- ❑ In some distributions there may be two or more mode of equal concentration which make the mode largely useless in such situations.
- ❑ It deliberately excludes arithmetical precision as its aim is to present a really typical measure.
- ❑ Mode is particularly useful in age and growth studies of fishes.

6. Measures of Dispersion

Dispersion refers to the extent to which numerical data is spread or dispersed around an average value. It is a measure of the variability of the data. In other words, dispersion helps us understand how the data points are distributed. The measure of dispersion quantifies the degree to which observations vary around the central tendency value. Unlike measures of central tendency, which describe the typical or average value of the data, measures of dispersion focus on how the observations are scattered around this central value. It's important to note that two sets of data may have the same mean, but the observations in one dataset may be widely scattered around the mean, while in another dataset, they may be more closely clustered. Therefore, measures of dispersion provide valuable insights into the variability or spread of the data. For example, consider two sets of data on length of fish in centimeters.

Set 1: 16 16 15 16 17 16

Set 2: 23 15 13 18 11 16

Both the above sets have the same mean value 16, although the pattern of individual observations is different in both the cases. In set 1, the observations congregate around the mean, whereas, in the second none of the observations have the value of the mean and are highly scattered. Thus in order to get the true picture of the data a measure of the central tendency has to be complemented by a suitable measure of dispersion.

Important Measures of Dispersion

1. Range
2. Mean Deviation

3. Variance
4. Standard Deviation

1. Range

- It is simply the difference between the highest value and the lowest value given in a data set.
- **Ungrouped data:** Range (R) = Highest Value – Lowest Value
- **Grouped data:** Range (R) = Highest class mark - Lower mark

Example

We have a dataset representing the heights of students in a classroom:

Height (in inches): 63, 65, 59, 72, 68, 67, 70, 64, 61, 66

To find the range in this dataset, we need to subtract the smallest value from the largest value.

Solution Here's how you can calculate it:

- **Arrange the data in ascending order:** 59, 61, 63, 64, 65, 66, 67, 68, 70, 72.
- **Identify the smallest and largest values:** Smallest value = 59 Largest value = 72.
- **Calculate the range:** Range = Largest value - Smallest value = 72 - 59, = 13 inches.

So, the range of heights in this dataset is 13 inches.

Merits	Demerits
• It is easy to understand. • It is easy to calculate. • It is rigidly defined. • It can be calculated from extreme values only. • In certain types of problems like quality control, weather forecasts, share price analysis, etc., range is most widely used.	• It is highly affected by extreme values. • In case of open-ended distributions, it is difficult to calculate. • It doesn't tell us much about distributions of items of the series. • It doesn't tell us much about distributions of items of the series. • It is very rarely used measure.

2. Mean Deviation (Average Deviation)

The arithmetic mean of the difference of each observation from a mean (ignoring sign) is known as the mean deviation (also called mean absolute deviation).

(i) Calculation for Ungrouped Data

Let $x_1, x_2, \ldots, x_k$ represent the values of the characteristic under study. Then the mean deviation is given by,

$$Mean\,Deviation = \frac{|x_1 - \overline{x}| + |x_2 - \overline{x}| + \ldots + |x_n - \overline{x}|}{n}$$

$$Mean\,Deviation = \frac{\sum |x - \overline{x}|}{n}$$

Example

2, 4, 8, 10. Calculate the mean deviation.

Solution

We find the mean of the dataset i.e. $\frac{2+4+8+10}{4} = 6$.

$$Mean\,Deviation = \frac{|x_1 - \overline{x}| + |x_2 - \overline{x}| + \ldots + |x_n - \overline{x}|}{n}$$

We then subtract each value in the dataset with the mean, get their absolute values i.e. |2-6| = 4, |4-6| = 2, |8-6| = 2, |10-6| = 4

And add them i.e. 4+2+2+4 = 12.

Finally, we divide this sum by the total number of values in the dataset (4) that will give us the mean deviation. The answer is 12/4 = 3.

(ii) Calculation for Grouped Data

Suppose the data are given in the form of a frequency distribution in k classes. Let $x_1, x_2, \ldots, x_k$ represent the midpoints of the 1st, 2nd, ..., kth class, and let $f_1, f_2, \ldots, f_k$ represent their frequencies respectively. Then the mean deviation is given by

$$Mean\ Deviation\ = \frac{|x_1 - \bar{x}|f_1 + |x_2 - \bar{x}|f_2 + \cdots + |x_k - \bar{x}|f_k}{n}$$

$$Mean\ Deviation\ = \frac{\sum |x - \bar{x}|f}{n}$$

Example

Class Interval	10 - 16	17 - 23	24 - 30	31 - 37	38 - 44	45 - 51
Frequency	8	10	23	29	18	18

Solution

Mean Deviation about mean

$MD = \frac{\sum |x - \overline{x}| f}{n}$. Where $\overline{X}$ the arithmetic mean.

Class Interval	f	x	fx	$\|x - \bar{x}\|$	$f\|x - \bar{x}\|$
9.5 -16.5	8	13	104	19	152
16.5 - 23.5	10	20	200	12	120

Class Interval	f	x	fx	$\|x-\bar{x}\|$	$f\|x-\bar{x}\|$
23.5 - 30.5	23	27	621	5	115
30.5 -37.5	29	34	986	2	58
37.5 - 44.5	18	91	738	9	162
44.5 – 51.5	12	48	576	16	192
	$\sum$ f=100		$\sum$ fx=3225		$\sum f\|x-\bar{x}\| = 799$

$$\text{Mean } (\bar{x}) = 32.25$$

$$\text{MD} = \frac{\Sigma f|x-\bar{x}|}{n} = \frac{799}{100} = 7.99 = 8$$

Merits	Demerits
• It is simple to understand. • It is based on all the observations of a series. • It shown the dispersion, or scatter of the various items of a series from its central value. • It is not very much affected by the values of extreme items of a series. • It facilitates comparison between different items of a series. • It truly represents the average of deviations of the items of a series. • It has practical usefulness in the field of business and commerce.	• If average is in fractions, it is difficult to compute. • It is not capable of further algebraic treatment. • It violates the algebraic principle by ignoring the + and – signs while calculating the deviations of the different items from the central value of a series. • It is less useful measure for the higher study of Statistics because of its dependence on the absolute value of deviation from the mean. • As for mean, open end series cannot be taken for the true result.

3. Variance

Variance in statistics is a measure of the spread or dispersion of a set of data points. It quantifies the average squared difference between each data point and the mean of the dataset. In other words, it provides a numerical value to describe how far the data points in a dataset are from the mean. A higher variance indicates that the data points are more spread out, while a lower variance suggests that the data points are closer to the mean.

The variance of a population is calculated as the average of square of deviations of observations from the arithmetic mean.

Variance of a population usually denoted by 'σ^2' whereas variance of a sample is denoted by '$\mathbf{s^2}$'.

4. Standard Deviation

Standard deviation is a statistical measure of the amount of variation or dispersion in a set of values. It is the positive square root of the variance and is expressed in the same units as the data.

Standard deviation of population usually denoted by 'σ' whereas standard deviation of a sample is denoted by 's'.

(i) Calculation of Variance for Ungrouped Data

For Population,

$$\sigma^2 = \frac{\sum(x-m)^2}{N}$$

$$\sigma^2 = \frac{[\sum x^2 - \frac{(\sum x)^2}{N}]}{N}$$

$$\sigma^2 = \frac{[N\sum x^2 - (\sum x)^2]}{N^2}$$

In the above expression 'm' is the population mean.

For Sample,

$$s^2 = \frac{\sum(x-\bar{x})^2}{(n-1)}$$

$$\sigma^2 = \frac{[\sum x^2 - \frac{(\sum x)^2}{n}]}{n-1}$$

$$\sigma^2 = \frac{[n\sum x^2 - (\sum x)^2]}{n(n-1)}$$

In the above expression '$\bar{x}$' is the sample mean.

Example

Calculate the variance and standard deviation for the following sample data on length (cm) of fingerlings: 9, 5, 8, 9, 7, 4, 10, 8

Solution

Sl.No	1	2	3	4	5	6	**7**	**8**	**Total**
X	9	5	8	9	7	4	10	8	60
X^2	81	25	64	81	49	16	100	64	480

Number of observations, n = 8

$$\text{Variance } (s^2) = \frac{n\sum x^2 - (\sum x)^2}{n(n-1)}$$

$$= \frac{8(480)-(60)^2}{(8)(7)}$$

$$= \frac{240}{56}$$

$$= 4.2857$$

Standard deviation, $s = \sqrt{4.2857} = 2.07$

(ii) Calculation of Variance for Grouped Data

For population,

$$\sigma^2 = \frac{\sum f(x-m)^2}{N}$$

$$\sigma^2 = \frac{[N\sum fx^2 - (\sum fx)^2]}{N^2}$$

For sample,

$$s^2 = \frac{\sum f(x-\bar{x})^2}{(n-1)}$$

$$\sigma^2 = \frac{[n\sum fx^2 - (\sum fx)^2]}{n(n-1)}$$

Example

Calculate the variance for the following data:

Class intervals	Frequency
200 – 201	13
201 – 202	27
202 – 203	18
203 – 204	10
204 – 205	1
205 – 206	1

Solution

Class Interval	f	x	fx	fx²
200 – 201	13	200.5	2606.5	522603.25
201 – 202	27	201.5	5440.5	1096260.75
202 – 203	18	202.5	3645	738112.5
203 – 204	10	203.5	2035	414122.5
204 – 205	1	204.5	204.5	41820.25
205 – 206	1	205.5	205.5	42230.25
	∑f = 70		∑fx = 14137	∑fx² = 2855149.5

$$s^2 = \frac{1}{n-1}\left[\sum fx^2 - \frac{(\sum fx)^2}{n}\right]$$

$$s^2 = \frac{1}{70-1}\left[2855149.5 - \frac{(14137)^2}{70}\right]$$

$$s^2 = 1.179$$

Standard Deviation = $\sqrt{\text{Variance}}$

$$= \sqrt{1.179}$$

$$= 1.0858$$

Merits of standard deviation	Demerits of Standard Deviation
• It is rigidly defined, and its value is always definite, based on all the observations, and the actual signs of deviations are used • As it is based on arithmetic mean, it has all the merits of arithmetic mean. • It is the most important and widely used measure of dispersion. • It is possible for further algebraic treatment. • It is less affected by the fluctuations of sampling and hence stable. • It is the basis for measuring the coefficient of correlation and sampling.	• It is not easy to understand and it is difficult to calculate. • It gives more weight to extreme values because the values are squared up. • As it is an absolute measure of variability, it cannot be used for the purpose of comparison.

Relative Measure of Dispersion – Coefficient of Variation

- All the measures of dispersion have a unit attached to them. For example, standard deviation calculated from a data recorded in cm' will have units as cm' and so on.
- Comparison of dispersion using standard deviation in two data sets of data, may lead to fallacious results.
- Therefore, the variability of fish length recorded in centimeters and fish weight in grams cannot be compared using standard deviation, as they are in different units. In such situations a relative measure of dispersion is used.
- Coefficient of variation is independent of the units. It is particularly useful when comparing the variability of different datasets with different units or scales.
- It is calculated as the ratio of standard deviation to arithmetic mean and is expressed in percentage.

$$CV = \frac{Standard\ Deviation}{Arithmetic\ Mean} \times 100$$

7. Concepts of Normal Curve, Skewness and Kurtosis

1. Normal Curve

The normal curve, also known as a Gaussian distribution or bell curve, is a continuous probability distribution that is symmetric around its mean (average).

Properties of Normal Curve:

- In a normal distribution, the mean, median and mode are equal. (i.e., Mean = Median= Mode).
- The total area under the curve should be equal to 1.
- The normally distributed curve should be symmetric at the centre.
- There should be exactly half values to the right of the centre and exactly half values to the left of the centre.
- The normal distribution should be defined by the mean and standard deviation.
- The normal distribution curve must have only one peak. (i.e., Unimodal)
- The curve approaches the x-axis, but it never touches, and it extends farther away from the mean.

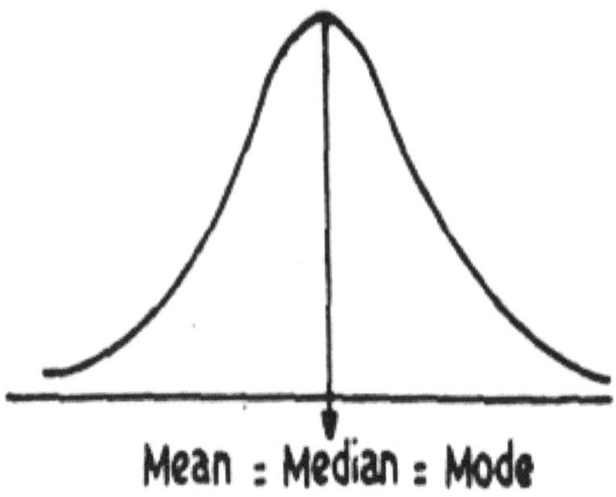

Normal Curve

2. Skewness:

- Even if two frequency distributions have similar means and variances, their frequency curves can still differ significantly from each other.
- Skewness measures the departure from symmetry in a distribution.
- It quantifies the extent of distortion from symmetry or the presence of asymmetry in a dataset.
- It serves as a representation of how much a distribution deviates from a normal distribution.

(i) Positively Skewed Curve:

- The frequency curve of distribution has longer tail to the right of the central maximum than to the left, the distribution is said to be skewed to the right or the positive skewness.
- **Positive Skewness:** Mean > Mode

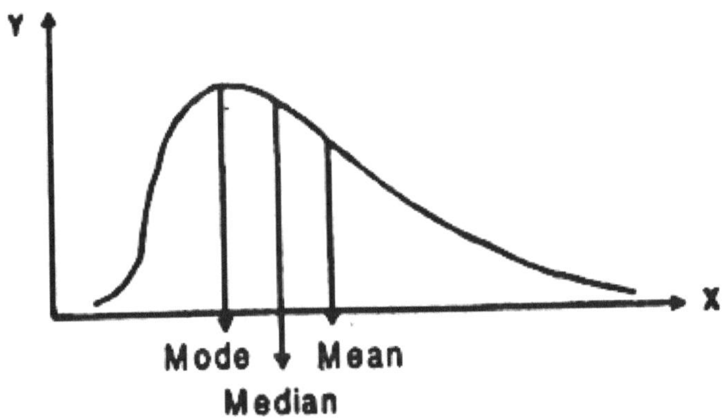

Positively Skewed Curve

(ii) Negatively Skewed Curve:

- The frequency curve of distribution has longer tail to the left of the central maximum than to the right, the distribution is said to be skewed to the left or the negative skewness.

Positive Skewness: Mean < Mode

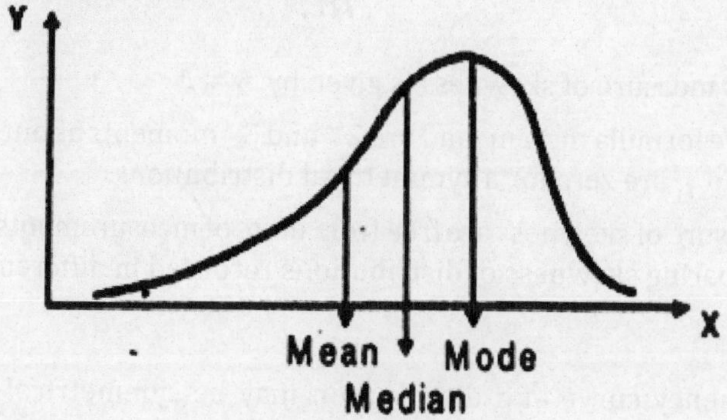

Negatively Skewed Curve

(iii) Measure of Skewness:

Skewness is quantified using a numerical value called the coefficient of skewness. The formula for the coefficient of skewness varies depending on the dataset and the specific skewness measure being used (there are different methods, based on mode, median and third moment about mean).

(a) Based on Mode:

The degree of skewness can be measured by the Pearson's coefficient of skewness which has the following formula.

$$Coefficient\ of\ Skewness = \frac{(Mean - Mode)}{Standard\ Deviation}$$

(b) Based on Median

In some frequency distribution, it may be difficult to calculate mode accurately. In such cases the coefficient of skewness which is based on median can be used.

$$Coefficient\ of\ Skewness = \frac{3(Mean - Median)}{Standard\ Deviation}$$

- This coefficient is zero for symmetrical distribution, positive for positively skewed distribution and negative for negatively skewed distribution.

(c) Based on third moment about mean:

An important and commonly used measure of skewness, which is based on the third moment about mean is given by,

$$\beta_1 = \frac{m_3^{\ 2}}{m_2^{\ 3}}$$

Sometimes measure of skewness is given by $\gamma_1 = \sqrt{\beta_1}$

In the above formula m_2 & m_3 and are 2nd and 3rd moments about the mean. The values of β_1 and γ_1 are zero for a symmetrical distributions.

These measure of skewness are free from units of measurements and therefore useful in comparing skewness of distributions recorded in different units.

3. Kurtosis:

- Frequency curve of a distribution may be symmetrical it can still be different in structure from the normal curve.
- For instance, it may exhibit greater or lesser peakedness compared to the normal curve
- Kurtosis is the degree of peakedness of a frequency curve as compared to the normal curve.
- A curve of relatively higher peakedness than the normal curve is called leptokurtic, whereas the curve having a flat top is called platykurtic.
- The normal curve which is neither very peaked nor flat-topped is called mesokurtic.

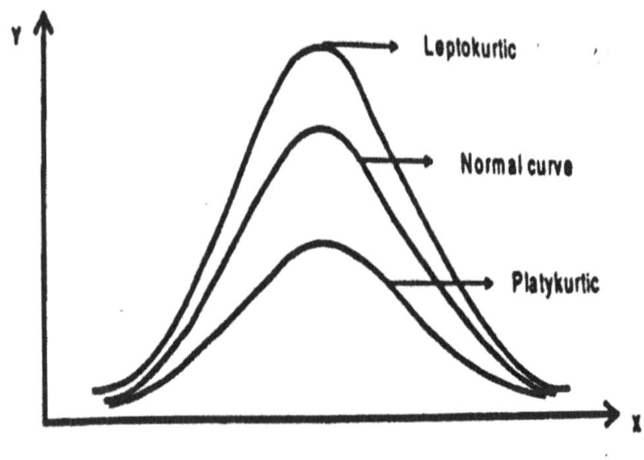

Kurtosis

Measure of Kurtosis:

Kurtosis is measured using a numerical value known as the kurtosis coefficient. The formula for calculating kurtosis may vary depending on the specific method used, but it essentially involves comparing the distribution to a normal distribution.

Based on the 4th moment about the mean

$$\beta_2 = \frac{m_4}{{m_2}^2}$$

- In the above formula m_4 is the 4th moment about the mean and M_2 is the 2nd moment about the mean.
- For normal distribution, $\beta_2 = 3$. Hence, sometimes measure of kurtosis is given by $\gamma_2 = \beta_2 - 3$.
- In normal distribution γ_2 is zero. In leptokurtic γ_2 value is positive and in platykurtic the γ_2 value is negative.
- The degree of kurtosis is indicated by the magnitude of this coefficient.

Example

Calculate the first four moments of the following distribution about the mean. Also find the measure of skewness and kurtosis.

X	0	1	2	3	4	5	6	7	8
f	1	8	28	56	70	56	28	8	1

Solution

X	f	di= $(x_i$-A)	$f_i d_i$	$f_i(x_i - \bar{x})2$	$f_i (x_i - \bar{x})3$	fi $(x_i - \bar{x})4$
0	1	-4	-4	16	-64	256
1	8	-3	-24	72	-216	648
2	28	-2	-56	112	-224	448
3	56	-1	-56	56	-56	56
4	70	0	0	0	0	0
5	56	1	56	56	56	56
6	28	2	56	112	224	448
7	8	3	24	72	216	648
8	1	4	1	16	64	256
Total	256		0	512	0	2816

$$Mean = A + \sum_{i=1}^{n} \frac{fidi}{N}$$

= 4+0/256

= 4

1st central moment: $\mu_1 = \frac{1}{N}\sum f_i\left(x_i - \bar{x}\right)^1$

= 1/256 x 512

= 0

2nd central moment:

$$\mu_2 = \frac{1}{N}\sum f_i(x_i - \bar{x})^2$$

= 1/256 x 512

= 2

3rd central moment:

$$\mu_1 = \frac{1}{N}\sum f_i\left(x_i - \overline{x}\right)^3$$

= 1/256 x 0

= 0

4th central moment:

$$\mu_1 = \frac{1}{N}\sum f_i\left(x_i - \overline{x}\right)^4$$

= 1/256 x 2816

= 0

Skewness= $\mu_3^2/\ \mu_2^3$= 0

Kurtosis = $\mu_4/\ \mu_2^2$= 11/4= 2.75

Chapter-3

Probability & Distributions

8. Probability – Definition and Concepts

Probability denotes the possibility of the outcome of any random event. It is a branch of mathematics that deals with the likelihood or chance of events occurring. Probability provides a way to quantify uncertainty and make informed decisions based on uncertain outcomes. Probability had its beginning with games of chance, such as the tossing of a coin, throwing a dice, drawing a card, etc., in the 17th century. It is used in various fields, including statistics, finance, engineering, biology, gaming, and more. It is fundamental in areas like risk assessment, decision-making under uncertainty, and experimental design. Probability forms the basis of statistical inference, which involves drawing conclusions about populations based on data from a sample. Techniques like hypothesis testing and confidence intervals rely heavily on probability theory.

Terminologies

Deterministic experiment

- If the same results are obtained when an experiment is repeated under same conditions, such an experiment is called deterministic experiment.
- For example, for a perfect gas, PV = constant, provided temperature is constant.
- Water activity for pure water is 1.00.
- The same result will be obtained whenever the experiment is repeated.
- Thus, the results of the deterministic experiment can be predicted with certainty.

Random Experiment

- The experiment which does not yield the same result when repeated under the same conditions is called a random experiment.
- In such an experiment it is impossible to predict the result with certainty.
- For example, tossing a coin can either yield head or tail.
- It is not possible to predict the outcome with certainty in advance.

Event / Simple event

- Every distinct outcome of a random experiment is called a simple event or an outcome.
- For example, in tossing a coin, head is one outcome and tail is another outcome.
- Hence, head and tail are the two events in the tossing of a coin experiment.
- In throwing a die, getting number 1 on top is one event, similarly getting 2, 3, 4, 5, 6 are the other events.

Sample space

- The totality or collection of all possible outcomes of a random experiment is called sample space. It is denoted by S.
- For example, in tossing of a coin experiment sample space consists of two events Head (H) and Tail (T).
- It is usually written as S = (H, T)

Exhaustive Events

- The total number of possible outcomes in any random experiment or a trial is known as exhaustive events or exhaustive cases.
- For example, in rolling a die, there are 6 exhaustive cases.

Compound Event

- A compound event is the one which consists of two or more simple events.
- For example, getting an even number in rolling a die is a compound event.

Equally likely Events or Outcomes (Equal outcomes)

- The outcomes are said to be equally likely when there is no reason to expect anyone rather than the other.
- For example, in tossing of a coin experiment, either head or tail may appear, so that both the outcomes are equally likely.

Mutually Exclusive Events

- To events A and B are said to be mutually exclusive if the occurrence of one event precludes the occurrence of another, i.e., both the events cannot happen simultaneously. In other words, A and B have no common outcomes.

- In the tossing of a coin experiment, head and tail are mutually exclusive events, as they cannot happen simultaneously.
- In an experiment on rolling a die, events of getting number 1 on top and events of getting number 2 on top are mutually exclusive, as numbers 1 and 2 cannot appear on the top of a die simultaneously.

Definition of probability

Definition 1: Classical or mathematical definition

Suppose an event E happens in 'm' different ways (outcomes) out of 'n' different exhaustive, mutually exclusive and equally like ways, then the probability of occurrence of an event denoted by P(E) is given by

$$P = P(E) = \frac{\text{Number of favorable ways of } E}{\text{Total number of equally likely ways}}$$

$$= \frac{m}{n}$$

- Probability of an event is a non-negative number which lies between 0 and 1, symbolically. $0 \leq p \leq 1$.

$$q = \frac{n-m}{n} = 1 - \frac{m}{n} = 1 - p$$

- So that, p + q = 1.
- *i.e.* the sum of the probabilities of occurrence and non-occurrence of an event is equal to 1.

Definition 2: Relative frequency or empirical definition of probability

- Classical definition of probability defined earlier assumptions that outcomes are equally likely, i.e. total number of outcomes are known and finite.
- When these assumptions are not met with, it is not possible to compute the probability of an event using the classical definition.
- In order to overcome the above limitations, a new approach called relative frequency concept of probability is adopted.
- According to this concept, the probability of the occurrence of an event E is the limiting value of the ratio of the frequency of occurrence of the event to the total number of outcomes.

For instance, if an experiment was repeated n times under the same conditions and if an event 'E' has occurred 'f' times, then the estimate of the probability of an event 'E' as the number of trials 'n' increases indefinitely is given by,

$$P(E) = \lim_{n \to \infty} \frac{f}{n}$$

(i) Independent Events:

- The events A and B are said to be independent if one occurrence does not depend on the occurrence or non-occurrence of the other. For instance, when a coin is tossed two times, the result of the second throw would no way be affected by the result of the first throw.
- Independent events can have common outcomes.
- If A and B are two independent events,

$$P(AB) \text{ or } P(A \cap B) = P(B).\ P(A)$$

(ii) Conditional Probability:

Let A and B be any two events. The conditional probability of event A, given that event B has happened, is denoted by P (A/B). Similarly, the conditional probability of B, given that event A has happened, is given by P (B/A).

$P(A/B) = P(A \cap B) / P(B)$

$P(B/A) = P(A \cap B) / P(A)$

(iii) Addition Theorem:

- Let A and B are two events with respective probabilities P (A) and P (B).
- The probability of occurrence of at least one of these two events denoted by P (A+B) is given by,

$P(A+B) = P(A) + P(B) - P(AB)$,

Where P (AB) is the probability of simultaneous occurrence of A and B.

Corollary: If events A and B are equally exclusive then,

$P(A+B) = P(A) + P(B)$

It is because of P (AB) = 0 when A and B are mutually exclusive.

(iv) Multiplication Theorem

- Let A and B be two events with probability P (A) and P (B), respectively.
- Let P (B/A) denote the conditional probability of the event B, given that event A has happened, P (A/B) the conditional probability of event A, given that event B has happened. Then, the probability of occurrence of both the events A and B denoted by P (AB) is given by,

$P(AB) = P(A) . P(B/A)$

$\qquad = P(B) . P(A/B)$

- If events A and B are independent, then

$$P(AB) = P(A) . P(B)$$
$$= P(B) . P(A)$$

Example-1

A coin is thrown 3 times .what is the probability that atleast one head is obtained?

Solution

Sample space = [HHH, HHT, HTH, THH, TTH, THT, HTT, TTT]

Total number of ways = 2 × 2 × 2 = 8.

Favorite Cases = 7

$P(A) = \frac{7}{8}$

Example-2

In a composite fish culture experiment, fingerlings of 6 species of fish namely, rohu, catla, mrigal, common cap, silver carp and grass carp, were stocked in the ratio of 1:1:1:2.5:3:1.5 respectively. A fingerling is subsequently drawn, what is the probability that it is of catla?

Solution

Fingerlings of rohu, catla, mrigal, common carp, silver carp and grass carp are stocked in the ratio of 1:1:1:2.5:3:1.5 respectively.

Thus out of 10 fingerlings we have 1 fingerling of catla.

Hence, the probability that the fingerling drawn is of catla $= \frac{1}{10} = 0.10$

Example-3

A problem is given to three fisheries students P, Q, R whose respective chances of solving it are 2/7, 4/7, 4/9 respectively. What is the probability that the problem is solved?

Solution

Probability of the problem getting solved = 1 – (Probability of none of them solving the problem)

$$P(P) = \frac{2}{7}; P(\bar{P}) = 1 - \frac{2}{7} = \frac{5}{7}$$

$$P(Q) = \frac{4}{7}; P(\bar{Q}) = 1 - \frac{4}{7} = \frac{3}{7}$$

$$P(R) = \frac{4}{9}; P(\bar{R}) = 1 - \frac{4}{9} = \frac{5}{9}$$

Example-4

The basket contains 6 catla and 3 rohu fishes. A sample of 3 fishes are taken at random without replacement from the basket. Calculate the probability of the following

(i) The sample contains 3 catla

(ii) The sample contains 2 catla and 1 rohu

(iii) The sample contains atleast 2 catla

Solution

- Total no. of fishes in the basket = 9

Catla = 6; Rohu = 3

(I) Sample contains 3 catla,

$$\frac{9\,C_3}{9C_3}\,;\ \mathrm{n}\,C_r = \frac{n!}{(n-r)!r!} = \frac{6!}{(6-3)!3!} \Big/ \frac{9!}{(9-3)!3!}$$

$$= (6\times5\times4\times3\times2\times1)/3!3! \Big/ \frac{9\times8\times7\times6\times5\times4\times3\times2\times1}{6!3!}$$

$$= \frac{20}{3\times2\times1} \Big/ \frac{504}{3\times2\times1} \quad = \frac{120}{6} \Big/ \frac{504}{6}$$

$$= \frac{120}{504}$$

$$= 0.23$$

(ii) Sample contain 2 catla and 1 rohu

$$\frac{6c_2\times3c_1}{9c_3} \quad = \frac{45}{85}$$

$$= 0.536$$

(iii) Sample contains atleast 2 catla

$$\frac{6c_2\times3c_1}{9c_3} + \frac{6c_3}{9c_3} \quad = 0.535 + 0.238$$

$$= 0.773$$

Example-5

In a certain district 45 % of the fish farmers practice composite fish culture of rohu, catla and mrigal. 35 % of fish farmers follow monoculture of rohu in their farm. Find the probability that randomly selected fish farmer follows atleast of the practices.

Solution

Let A and B be the events are:

A : Practicing composite fish culture

B : Practicing Monoculture

P(A) : 45 % = 45 / 100

P(B) : 35 % = 35 / 100

P(A . B) = 20 % = 20 / 100

P(A+B) = P(A) + P(B) – P(AB)

= 45 / 100 + 35 / 100 – 20 / 100

= 60 / 100

= 0.6

9. Counting Techniques – Permutations, Combinations & Bayes' Theorem

Factorials

- For any given positive integer 'n' the product of all the whole numbers from 'n' down through 1 is called n factorial and is written as n !
- For example, $5! = 5 \times 4 \times 3 \times 2 \times 1 = 120$
- By definition, $0 ! = 1$
- And in general, $n ! = n (n-1) !$

  $= n (n-1) (n-2) !$ And so on
- Factorials are useful in finding the number of ways objects can be arranged in a line. For example, suppose there are 3 containers of culture media, each of which is inoculated with a different organism. These culture media can be placed in a line on a platform in $3 ! = 6$ ways. If the 3 media are designated as a, b, and c, then 6 arrangements are abc, bca, cab, cba, bac, acb.

Permutations

- Permutation is an ordered arrangement of objects.
- The 6 arrangements of culture media indicated above are the possible permutations of 3 objects taken 3 at a time.
- Many times there may be more objects than the positions to fill.

- Suppose there are 3 culture media and only two positions are available on the platform. Then the number of possible permutations taken two at a time is 3 × 2 = 6 because there are 3 objects to fill the first position, and once that position is filled, there are 2 objects only left to fill the second position.
- The 6 permutations are ab, ba, ac, ca, bc, cb.
- In general, the number of possible permutations of n objects taken r at a time denoted by nPr given by
- nPr = n (n-1) (n-2) ..... (n-r+1)

$$= \frac{n!}{(n-r)!}$$

Combination

- Combination is an arrangement of objects without considering their orders. As indicated above the 6 permutations of 3 objects taken 2 at time are ab, ba, ac, ca, bc, cb. It may be noticed here that ab and ba, ac and ca, bc and cb are the same except for the order in which the letters occur. Thus, when the order is ignored, there will be only 3 combinations.
- This is because there will be two permutations for each combination. In general, there will be r! permutations for each combination of n objects taken r at a time.
- Where nCr refers to number of combinations of n objects taken r at a time.
- Hence, $nCr = \frac{nPr}{r!}$

$$= \frac{n!}{(n-r)!\,r!}$$

Bayes Theorem

- Bayes' theorem describes the probability of the occurrence of an event related to any condition. It is also known as the formula for the probability of "causes."
- Bayes' theorem is concerned with determining the probability of an event given certain new information that has been acquired from a sample, a survey, or an experiment.
- For example, a sample output of 4 defectives in 25 trials (event A) can be used to estimate the probability that a machine is not working properly (event Ei).
- Let $E_1, E_2, ..., E_n$ be a set of events associated with a sample space S, where all the events $E_1, E_2, ..., E_n$ have a nonzero probability of occurrence, and they form a partition of S. Let A be any event associated with S, then according to Bayes' theorem,

$$P\,(Ei/A) = \frac{P\,(Ei)\,P\,(A/Ei)}{\sum_{i=k}^{k} P\,(Ei)\,P\,(A/Ei)}, \text{ for any i} = 1, 2, 3, ..., k$$

The following terminologies are also used when Bayes' theorem is applied:

- Hypotheses: The events E1, E2, ..., En are called hypotheses.
- Priori Probability: The probability P(Ei) is considered the priori probability of hypothesis Ei.
- Likelihoods Probability: The probability P(A/Ei), i = 1, 2, 3, ..., k, are called "likelihoods."
- Posteriori Probability: The probability P(Ei/A) is considered the posteriori probability of hypothesis Ei.

Example

In a coastal region, there are two fishing communities, A and B. Community A catches mackerel 60% of the time, while Community B catches tuna 80% of the time. Mackerel constitutes 40% of the total fish population in the region. If a randomly selected fish is found to be a mackerel, what is the probability that it was caught by Community A?

Solution

Let A be the event that the fish was caught by Community A, B be the event that the fish was caught by Community B and M be the event that the fish is a mackerel

To find P(A|M), the probability that the fish was caught by Community A is a mackerel.

According to Bayes' theorem:

$$P(A/M) = \frac{P(M/A) x\, P(A)}{P(M)}$$

From the problem we know the following,

- P(M|A) = 0.60 (the probability of catching a mackerel given that it's from Community A)
- P(A) = 0.50 (assuming equal chances for each community)
- P(M) = P(M|A) x P(A) + P(M|B) x P(B)

Substituting the values:

$$P(A/M) = \frac{0.6 x 0.5}{P(M/A) x 0.5 + P(M/B)\ x 0.5}$$

$$= \frac{0.6 x 0.5}{0.6 x 0.5 + 0.2\ x 0.5}$$

$$= \frac{0.3}{0.4}$$

So, the probability that the fish caught by Community A is a mackerel is 0.75.

10. Probability Distribution

Random Variable

- In probability, a real-valued function defined over the sample space of a random experiment is called a random variable.
- That is, the values of the random variable correspond to the outcomes of the random experiment.
- Random variables could be either discrete or continuous.
- For example, let X be a random variable connected with the outcome of a random experiment. Consider an experiment of tossing a coin twice. Then the experiment has the outcomes HH, HT, TH, and TT, where H and T refer to head and tail, respectively. If the number of heads is considered as a random variable X, then its value corresponding to different outcomes is as follows:

Outcome	HH	HT	TH
Value of X	2	1	1

- Thus random variable is a function that assigns a real number to each outcome in a sample of a random experiment.

Probability Distribution:

- Probability distribution yields the possible outcomes for any random event. It is also defined based on the underlying sample space as a set of possible outcomes of any random experiment.
- Probability distribution is correlated to a relative frequency distribution with probabilities replacing relative frequencies.

- Thus probability distribution can be regarded as theoretical or limiting forms of relative frequency distributions, when the number of observations made is very large.
- Hence, probability distributions can be considered as distributions of populations, whereas relative frequency distributions are distributions of samples drawn from these populations.

Types of Distributions

(a) Discrete Probability Distribution (Probability Mass Function)

- Discrete random variable
- Listing of all possible values with respective probabilities of occurrence
- As discrete random variable can take only a finite number of values or countable infinite number of values, it is possible to list all the values with the corresponding probabilities
- Binomial and Poisson Distributions come under this distribution
- Example: The distribution of outcomes when rolling a fair six-sided die

(b) Continuous Probability Distribution (Probability Density Function)

- Continuous random variable
- It is not meaningful to assign probabilities to individual points, hence the probability of a random variable falling within a given interval is considered
- The Normal Distribution falls under this category of distribution
- Example: The distribution of heights of a population

1. Binomial Distribution or Bernoulli Distribution

- Binomial distribution is a discrete distribution.
- When an event (character) can occur in one of the two possible ways.
- This experiment is repeated 'n' independent times.
- **For example:**
    - ☆ Finding the quantity of raw and used materials while managing a fishery
    - ☆ Finding the number of male and female employees in a fisheries-related organization
    - ☆ Taking a survey of positive and negative reviews from the public for a specific fishery or fishing area
- Let p be the probability of success and q be the probability of failure.
- Let us assume that x outcomes are success and the remaining n-x outcomes are failures.

The mathematical description of the Binomial distribution is as follows:

$$P(X) = nC_x p^x q^{n-x} \;;\; x = 0, 1, 2, \ldots\ldots n$$

$$P(X) = \frac{n!\, p^x q^{n-x}}{x!\,(n-x)!} \;;\; x = 0, 1, 2, \ldots n$$

- Important Properties of the binomial distribution:
  - ☆ Mean of the binomial distribution = np
  - ☆ Variance of the binomial distribution = npq
  - ☆ Standard deviation = $\sqrt{npq}$
  - ☆ For p = q = ½, it is symmetrical; p < ½, it is positively skewed; p > ½, it is negatively skewed

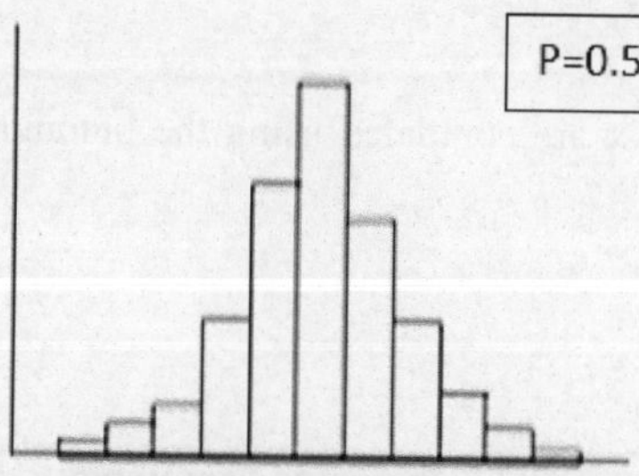

Symmetric distribution

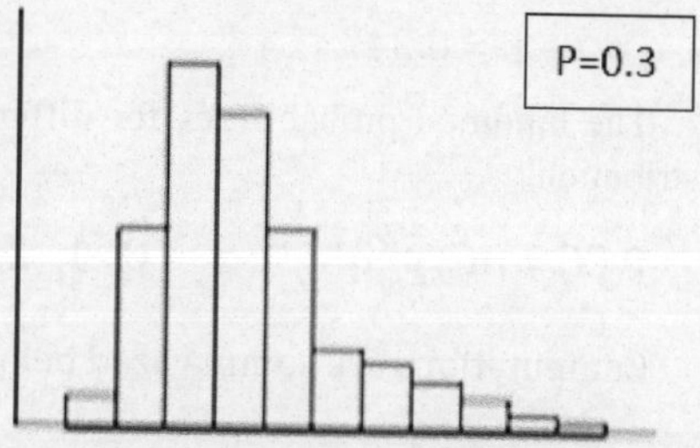

Positively skewed distribution

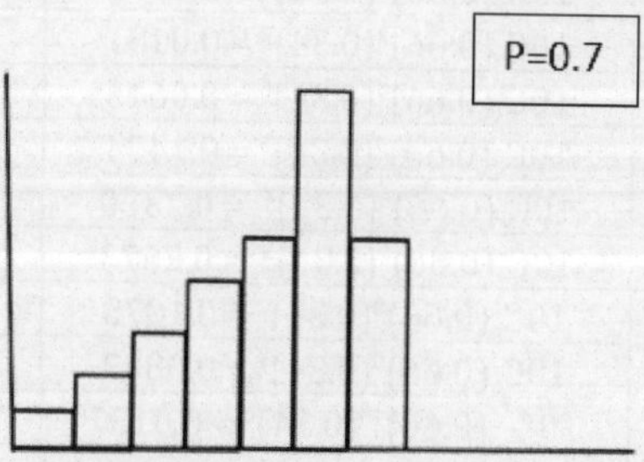

Negatively skewed distribution

Fitting of Binomial Distribution

- Binomial distribution is fitted by estimating p from the observed data.
- As the mean of the binomial distribution is '*np*', dividing mean by 'n' will give '*p*'.
- Once 'p' and 'n' are known, binomial probabilities and the expected frequencies can be computed.

Example

The number of sets of Catla which responded in induced breeding out of 10 sets tried per experiment was noted. A total of 100 such experiments were conducted in DIFST. The results are summarized in the following frequency distribution table.

No. of responded (x)	0	1	2	3	4	5	6	7	8	9	10
Frequency	1	1	1	2	4	12	22	27	19	9	2

Fit the binomial distribution

Solution

Here n = 10, $f = N = 100$

np = Mean = $\frac{\sum fx}{\sum f} = \frac{659}{100} = 6.59$

Therefore, p = Mean/n = 0.659 (approximately 0.66)

q = 1 – p = 0.34

The binomial probabilities for different values of x are computed using the binomial distribution.

P (X) = $nC_x p^x q^{n-x}$; $x = 0, 1, 2, \ldots$ n

Computations are summarized below:

x	f	fx	P (x)	Expected Frequency (N P (x))
0	1	0	10C0(0.66)0(0.34)10 = 0	0
1	1	1	$10C_1(0.66)^1(0.34)^9 = 0.0004$	0.04
2	1	2	$10C_2(0.66)^2(0.34)^8 = 0.0035$	0.35
3	2	6	$10C_3(0.66)^3(0.34)^7 = 0.0181$	1.81
4	4	16	$10C_4(0.66)^4(0.34)^6 = 0.0615$	6.15
5	12	60	$10C_5(0.66)^5(0.34)^5 = 0.1434$	14.34
6	22	132	$10C_6(0.66)^6(0.34)^4 = 0.2320$	23.20
7	2	189	$10C_7(0.66)^7(0.34)^3 = 0.2573$	25.73
8	19	152	$10C_8(0.66)^8(0.34)^2 = 0.1873$	18.73
9	9	81	$10C_9(0.66)^9(0.34)^1 = 0.0807$	8.07
10	2	20	$10C_{10}(0.66)^{10}(0.34)^0 = 0.0157$	1.57
	$\sum f$=100	$\sum fx$=659		

2. Poisson Distribution

- Poisson distribution is another discrete probability distribution which has frequent applications in faunal sampling operations where the character or variate under study is the number of animals or species per unit of observation.
- In practice, if the count data represent the number of rare events occurring within a given unit of time or space, the distribution of these counts can

be described by the Poisson distribution.

- In Poisson distribution, it is assumed that rare events occur randomly and independently.
- **If 'p' the probability of occurrence of an event is very small and 'n' the number of trials is very large such that np is constant, then Binomial distribution tends to a Poisson distribution.**
- Some examples of Poisson variable are number of ships arriving in a harbour per hour, number of animals per square of plankton species, the number of machines breaking down daily in a fish processing plant.
- A random variable X is said to follow Poisson distribution if the probability mass function is given by,

$$P\left(X=x\right) = \frac{e^{-\lambda}\lambda^{x}}{x!}; x = 0, 1, 2,......$$

- Where 'e' is the base of natural logarithm having a value of 2.7183, m is the mean of the distribution.
- If 'λ' is known, this distribution can be completely determined and is called parameter of the distribution.
- An important characteristic of the Poisson distribution is that its variance is equal to the mean of the distribution.
- The Poisson distribution is positively skewed. However, as λ (=np, where n is large) increase, it will resemble normal distribution.
- As variance equal mean in Poisson distribution, the ratio of s^2/m can be used to determine whether the variable under study is randomly distributed or over dispersed.
- Theoretically if this ratio is greater than 1, the population is over dispersed and Poisson distribution will not be suitable to describe this population.

Example

In a study conducted to analyze the catch rates of a specific fish species in a particular fishing area, a fishery biologist recorded the number of fish caught per hour over a period of 50 hours. The data collected is as follows:

Number of fish caught per hour	0	1	2	3	4
Frequency	109	65	22	3	1

Fit a Poisson distribution to following data.

Solution

Given, n =200

$$\lambda = \text{Mean } \frac{\sum fx}{\sum f} = \frac{122}{200} = 0.61$$

Number of fish caught per hour (x)	Frequency (f)	fx	P (x)	Expected Frequency (N P (x))
0	109	0	$\frac{e^{-0.061}(0.061)^0}{0!}=0.54$	108.6
1	65	65	$\frac{e^{-0.061}(0.061)^1}{1!}=0.33$	66.2
2	22	44	$\frac{e^{-0.061}(0.061)^2}{2!}=0.104$	20.4
3	3	9	$\frac{e^{-0.061}(0.061)^3}{2!}=0.02$	4
4	1	4	$\frac{e^{-0.061}(0.061)^4}{2!}=0.001$	0.02
	∑f=200	∑fx=122		

3. Normal Distribution

- The normal distribution is one of the most important distributions in statistics. Its equation was first given by De Moivere in 1733. Later it was rediscovered and developed by Gauss in 1809 and by Laplace in 1812. Therefore, this distribution is sometimes referred to as Gaussian and Laplace distribution.
- The probability density function of a random variable having normal distribution is given by,

$$(x)=\frac{1}{\sigma\sqrt{2\pi}}e^{\frac{-(x-m)^2}{2\sigma^2}};-\infty<X<\infty$$

- Where 'm' and 'σ'are respectively the mean and standard deviation of the distribution. π and e are the constants whose values are the equal to 3.1416 and 2.7183 respectively.
- The graph of f(x) is a famous 'bell shaped' curve.
- Normal distribution can be completely identified if mean and standard deviation are known.
- The distribution will vary depending upon the values of m and σ.
- It is a continuous distribution and can theoretically assume a value from-∞ to + ∞.
- However, for all practical purpose the values lie in the range of ±3 standard deviation from the mean.

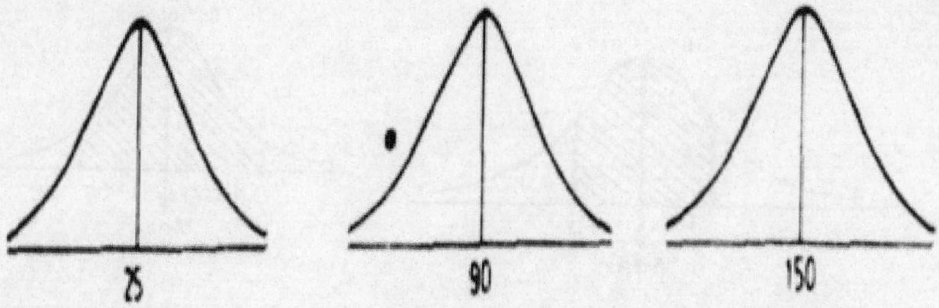

Distributions with same standard deviations but different means

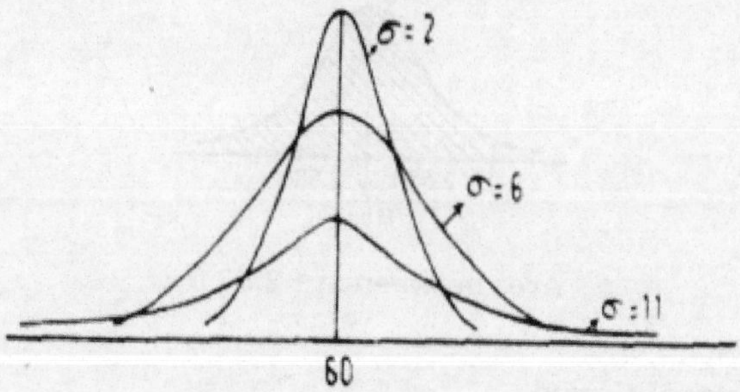

Distributions with same mean but different standard deviations

Properties of normal curve:

- ☆ It is a continuous, symmetrical, and bell-shaped curve.
- ☆ It is asymptotic; both tails extend to infinity, meaning they approach but never touch the base.
- ☆ The arithmetic mean, median, and mode coincide.
- ☆ The central position of the curve is described by the mean, and the spread of the curve is described by the standard deviation.
- ☆ The coefficient of skewness is zero, and the coefficient of kurtosis is 3.
- ☆ Mean plus or minus one standard deviation ($\mu \pm 1\sigma$) includes approximately 68% (specifically 68.27%) of the total frequency or area under the curve.
- ☆ Mean plus or minus 1.96 standard deviations ($\mu \pm 1.96\sigma$) includes approximately 95% of the total frequency or area under the curve. (This value corresponds to approximately 95% confidence interval in a normal distribution).
- ☆ Mean plus or minus 2.58 standard deviations ($\mu \pm 2.58\sigma$) includes approximately 99% of the total frequency or area under the curve. (This value corresponds to approximately 99% confidence interval in a normal distribution).

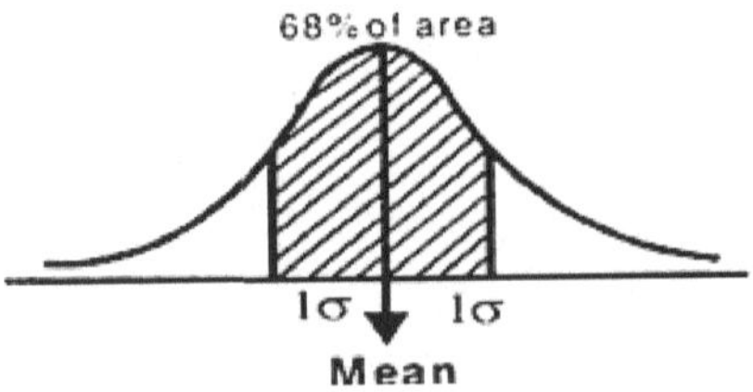

Area between m ± 1σ

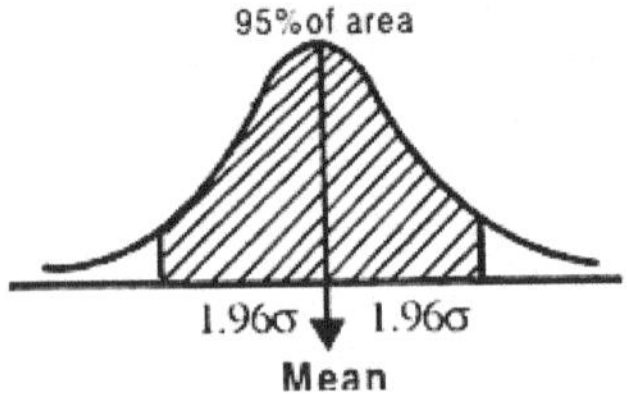

Area between m ± 1.96σ

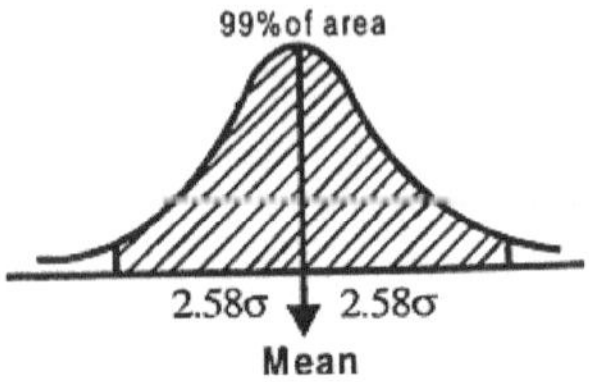

Area between m ± 2.58 σ

Area under Normal Curve

- The area bounded by the normal curve and the x-axis is 1.
- Quite frequently, the area under this curve that falls between two points on the x-axis, say x = a and x = b, is required. Instead of performing these calculations, tables are available for ready use.
- These tables provide the area under the normal curve, which has a mean of zero and a standard deviation of one (referred to as the standard normal curve).
- Hence, to utilize these tables, the normal variable X must be transformed into the standard normal variable Z using the following relation: $Z = (X - \mu) / \sigma$
- As the standard normal curve is symmetric about Z = 0, the area between Z = 0 and any negative Z value, say Z = -a, is equivalent to the area between Z = 0 and Z = +a.

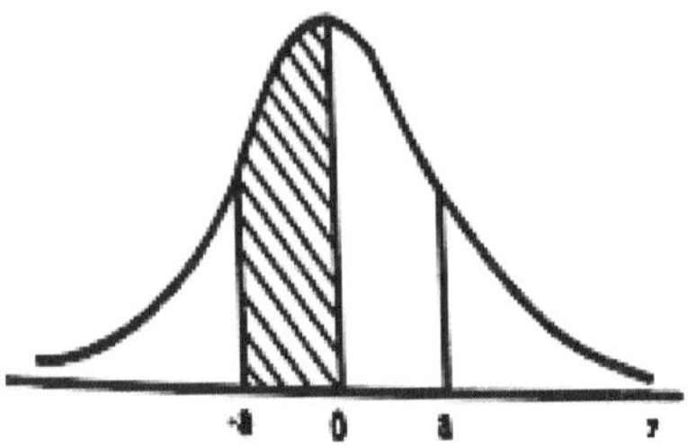

Standard Normal Curve

Different Forms of Area Tables:

- In fisher and Yates (1963) the area of normal curve is tabulated from Z to ∞.

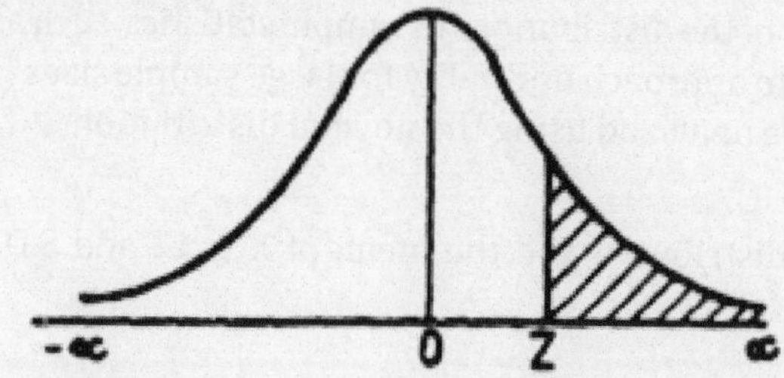

- In Spiegel (1981) the area of normal curve is tabulated from 0 to any positive value of Z.

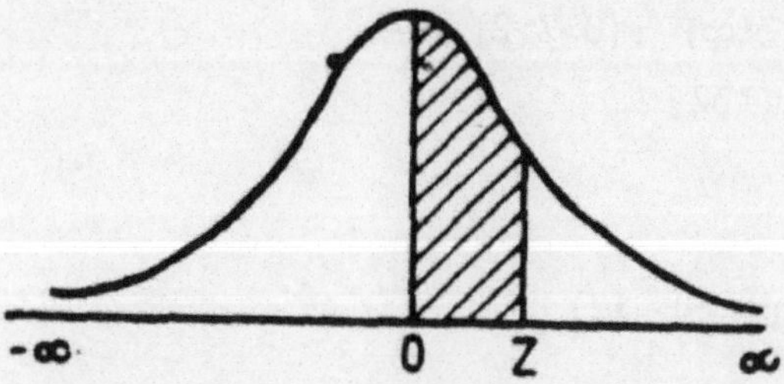

- In Woolf (1968) area of the normal curve is tabulated from - ∞ to Z.

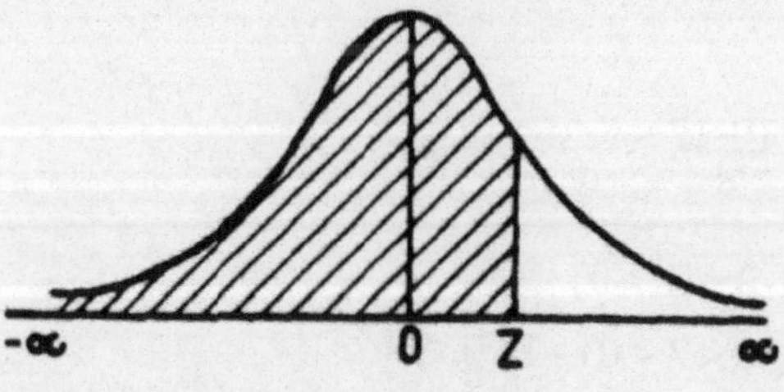

Importance of the Normal Distribution:

- ☆ Numerous continuous phenomena or characteristics such as fish length, weight, body depth, etc., are approximately normally distributed.
- ☆ Many of the discrete distributions occurring in practice, such as the Binomial, Poisson, etc., can be approximated by a normal distribution.
- ☆ Even when the variable is not normally distributed, it is possible to bring it to approximate normality through simple transformations such as square root, logarithmic, or arcsine transformations, etc.

- ☆ The normal distribution has simple and interesting mathematical properties.
- ☆ Many distributions provide the basis for statistical inference.
- ☆ Many of the distributions of sample statistics, such as the sample mean, tend to approach normality for large sample sizes (n), and hence they can be analyzed using the normal distribution

Example-1

X is normally distributed and the mean of X is 12 and S.D is 4. Find out the probability of the following:

P (X>20)

$$P(X>20) = P\left(\frac{x-\mu}{\sigma} > \frac{20+2}{4}\right)$$

P (Z>2) = P(0 <Z<∞) – P(0>Z>2)

= 0.5 – 0.4772

= 0.0228

X< 20

$$P(X<20) = P\left(\frac{X-\mu}{\sigma} < \frac{20-12}{4}\right)$$

P (Z<2) = P (0< Z < -∞) + P(0 <Z<2)

= 0.5 + 0.4772

= 0.9772

0 <X<12

$$P(0< X< 12) = P\left(\frac{0-12}{4} < \frac{X-\mu}{\sigma} < \frac{12-12}{4}\right)$$

= P (-3 < Z < 0)

= P (-3 < Z < 0) – P(0< Z< ∞)

Therefore, probability of (-3 < Z< 0) from the table is = 0.4986

P (0<X<12) = 0.9986 -0.5

= 0.4986

P(X> X') = 0.25

$$P\left(\frac{X-\mu}{\sigma} > \frac{X'-\mu}{\sigma}\right) = 0.25$$

P (Z> Z') = 0.25

P (0 < Z< ∞) – P (0< Z< Z') = 0.25

0.5 - P (0< Z< Z') = 0.25

P (0< Z< Z') = 0.5 – 0.25

Now, P (-∞ < Z < Z') – P (-∞ < Z <0) = 0.25

P (-∞ < Z < Z') – 0.5 = 0.25

P (-∞ < Z < Z') = 0.5 + 0.25

= 0.75

From the table Z value for 0.75; Probability is 0.6 + 0.07

Z' = 0.67

$$Z = \frac{X-\mu}{\sigma}$$

$$=> 0.67 = \frac{X-12}{4}$$

= 2.67 + 12

X = 14.68

Example-2

The mean length of a one old brood of catla is 30cm and SD is 2cm. a fish is caught at random, find the probability that its length is

(a) Between 30 and 32cm.

(b) Between 28 and 33 cm.

(c) Suppose it was decided to transfer all those having length greater than 31cm, what percent of fish is required to be transferred; assume lengths are randomly distributed.

Solution

Given, μ=30cm,σ=2cm

(a) 30<X<32

=P(0<X<32)

$$= P\left(\frac{30-30}{2} < \frac{X-\mu}{\sigma} < \frac{32-30}{2}\right)$$

=P(0<Z<1)

=0.3413

(b) 28<X<33

$$= P\left(\frac{28-30}{2} < \frac{X-\mu}{\sigma} < \frac{33-30}{2}\right)$$

$$= P\left(-\frac{2}{2} < Z < \frac{3}{2}\right)$$

$=P(-1<Z<1.5)$

$=P(-1<Z<0)+P(0<Z<1.5)$

$=P(0<Z<1)+P(0<Z<1.5)$

$=0.3413+P(-\infty<Z<1.5)-P(-\infty<Z<0)$

$=0.3413+0.93319-0.5=0.77444$

Example-3

The marks obtained by no. of students for certain subjects are assumed to be approximately normally distributed with mean values 65 & standard deviation of 5. If 3 students are taken at random from this set, what is probability that exactly 2 of them will have marks over 70?

Solution

Given,

$\mu=65$ $\sigma=5$ $N=3$

Here, N is very small so we can apply Normal distribution here,

$$P(X > 70) = P(\frac{X-\mu}{\sigma} > \frac{70-65}{5})$$

$=P(Z>1)$

$=P(-\infty<Z<\infty)-P(-\infty<Z<1)$

$=1-0.84134$ ....(Table Values)

$=0.1587$

But $p+q=1$

$\therefore q=1-0.1587$

$\therefore q=0.8413$

$3C_2\, p^2q^1=3\times(0.1587)^2\times(0.8413)^1$

$=0.0636$

Chapter-4

Inferential Statistics, Hypothesis & Test Statistics

11. Sampling Distribution

A sampling distribution refers to the probability distribution of a statistic obtained from selecting random samples from a given population. Also known as a finite-sample distribution, it represents the distribution of frequencies indicating how spread out various outcomes will be for a specific population.

How its work?

- Select a random sample of a specific size from a given population.
- Calculate a statistic for the sample, such as the mean, median, or standard deviation.
- Develop a frequency distribution of each sample statistic calculated in the previous step.
- Plot the frequency distribution of each sample statistic developed in the previous step. The resulting graph represents the sampling distribution.

Example

For example, consider an artificial population of fingerling lengths (in cm) consisting of 5 fingerlings: 3, 4, 7, 5, 8. It is decided to estimate the mean length of fingerlings from a sample of 2. There will be 10 different samples of 2 fingerlings each if sampling without replacement, and 25 samples with replacement that can be drawn from a population of size 5. The possible samples of size 2 when sampling with replacement are listed in Table 1.

Table 1. Sample Consisting Individuals

	1	2	3	4	5
1	1,1	1,2	1,3	1,4	1,5
2	2,1	2,2	2,3	2,4	2,5
3	3,1	3,2	3,3	3,4	3,5
4	4,1	4,2	4,3	4,4	4,5
5	5,1	5,2	5,3	5,4	5,5

Table 2. Sample Values

	3	4	7	5	8
3	3,3	3,4	3,7	3,5	3,8
4	4,3	4,4	4,7	4,5	4,8
7	7,3	7,4	7,7	7.5	7,8
5	5,3	5,4	5,7	4.0	5,8
8	8,3	8,4	8,7	8,5	8,8

Table 3. Sample Means

	3	4	7	5	8
3	3.0	3.5	5.0	4	5.5
4	3.5	4.0	5.5	4.5	6.0
7	5.0	5.5	7.0	6.0	7.5
5	4.0	4.5	6.0	5.0	6.5
8	5.5	6.0	7.5	6.5	8.0

From Table 3, it can be observed that the means computed for different samples are not the same but vary. Thus, there is a distribution of the means, which is called the 'sampling distribution' of means. Likewise, if variance is calculated for different samples, it is likely to vary from sample to sample, resulting in the sampling distribution of variance. Similarly, sampling distributions of other statistics such as standard deviation, median, etc., can be obtained.

Types of Sampling Distributions:

- **Sampling Distribution of the Mean:** This method depicts a normal distribution where the center is the mean of the sampling distribution. It represents the mean of the overall population. To reach this point, the researcher must determine the mean of each sample group and plot the individual data accordingly.
- **Sampling Distribution of Proportions:** This method involves selecting a sample from the overall population to determine the proportion of the sample. The mean of the proportions then represents the proportions of the larger group.
- **T-Distribution:** This type of sampling distribution is common in cases of small sample sizes or when there is limited information about the entire

population. T-distributions are utilized to make estimates about the mean and other statistical parameters.

Importance of Using a Sampling Distribution

- The sampling distribution gives us a smaller range of the possible outcomes or results from a large range of data.
- Sampling distributions help eliminate variability when gathering statistical data or conducting research.
- Sampling distributions make data easier to manage.
- Sampling distributions form the basis for statistical inference, allowing for inferences to be made about the entire population.

12. Standard Error

The mean and standard deviation of sampling distributions of statistics can be computed similarly to the cases of probability distributions based on individual observations. The standard deviation of the sampling distribution of a statistic is called the standard error. Generally, the standard error decreases as the sample size increases.

Standard Error of Sampling Distribution of Mean

- The standard error (SE) of mean of sample is given by

$$Standard\ Error\ of\ Mean = \left[\frac{\sigma}{\sqrt{n}}\right]\left[\sqrt{\frac{N-n}{N-1}}\right]$$

Where 'σ' is the standard deviation of population.

- If the population is much larger than the sample or if sampling is with replacement, then

$$Standard\ Error\ of\ Mean = \left[\frac{\sigma}{\sqrt{n}}\right]$$

- If the population standard deviation is not known, then sampling standard deviation 's' can be used in its place. When 'n' is large, then

$$Standard\ Error\ of\ Mean = \left[\frac{s}{\sqrt{n}}\right]$$

$$S = \sqrt{\frac{\left(\sum x-\bar{x}\right)^2}{(n-1)}}$$

Example

The students conduct a study in a lake and catch 20 fish. Their weights, measured in kilograms, are as follows: 2.3, 2.5, 2.2, 2.4, 2.1, 2.3, 2.6, 2.4, 2.7, 2.2, 2.5, 2.3, 2.4, 2.6, 2.3, 2.1, 2.5, 2.4, 2.6, and 2.2. How do we calculate the standard error for the mean weight of these fish?

Solution

Sample size $(n) = 20$

Calculate the Sample mean $(\bar{x}) = \frac{2.3+2.5+\cdots+2.2}{20} = \frac{48}{20} = 2.4$

Calculate the Variance (s^2)

$$S^2 = \frac{(2.3-2.41)^2+(2.5-2.41)^2+\cdots+(2.2-2.41)^2}{20-1}$$

$$= \frac{0.0169+0.0081+\cdots+0.03240.0169+0.0081+\cdots+0.0324}{19} = 0.0134$$

Standard Error $(SE) = \frac{\sqrt{0.0134}}{\sqrt{20}} \approx 0.026$

Standard error of sum of difference of means:

If $\bar{x}_1$ denotes the mean of a sample size n_1 drawn from a population with variance σ_1^2 and $\bar{x}_2$ denote the mean of a sample size n_1 drawn from a population with variance σ_2^2 then variance and standard error of the distribution of sum or difference of sample mean is given by,

$$\text{Variance of } (\overline{x_1} \pm \overline{x_2}) = \frac{\sigma_1^2}{n_1} + \frac{\sigma_2^2}{n_2}$$

$$\text{Standard error of } (\overline{x_1} \pm \overline{x_2}) = \sqrt{\frac{\sigma_1^2}{n_1} + \frac{\sigma_2^2}{n_2}}$$

Example

The students are conducting a study on fish lengths in two different areas of a river. In the first area, we catch 5 fish, and in the second area, we catch 7 fish. The weights of the fish in kilograms are as follows:

Area 1 (Sample 1): 2.3, 2.5, 2.2, 2.4, 2.1

Area 2 (Sample 2): 2.6, 2.4, 2.7, 2.2, 2.5, 2.3, 2.4

Solution

$n_1 = 5, n_2 = 7$

Calculate the Sample means ($\bar{x}_1$ and $\bar{x}_2$)

For sample 1: $\bar{x}_1 = \frac{2.6+2.5+2.2+2.4+2.1}{5} = \frac{11.5}{5} = 2.3$

For sample 2: $\bar{x}_2 = \frac{2.6+2.4+2.7+2.2+2.5+2.3+2.4}{7} = \frac{18.1}{7} = 2.586$

Calculate the sample variances $(\sigma_1{}^2 \text{ and } \sigma_2{}^2)$

For sample 1: $\sigma_1{}^2 = \frac{(2.3-2.3)^2+(2.5-2.3)^2+\cdots+(2.1-2.3)^2}{5}$

$$= \frac{0+0.04+\cdots+0.04}{5} = \frac{0.16}{5} = 0.032$$

For sample 2: $\sigma_2{}^2 = \frac{(2.6-2.586)^2+(2.4-2.586)^2+\cdots+(2.4-2.586)^2}{7}$

$$= \frac{0.000256+0.007236+\cdots+0.007236}{7} = \frac{0.0473}{7} = 0.00675$$

Calculate the Standard Error (SE):

$$\text{SE} = \sqrt{\frac{0.032}{5} + \frac{0.00675}{7}} = \sqrt{0.0064 + 0.000964} = \sqrt{0.007364} = 0.0858$$

Standard error of a sampling distribution of proportion:

Let P denote the population proportion and Q = 1 – P. If p denotes sample proportion based on a sample size n, then the standard error of p is given by,

$$\textit{Standard Error} = \frac{PQ}{n}$$

If p_1 and p_2 denotes the sample proportions based on samples of size n_1 and n_1 respectively from the populations with proportion of P_1 and P_1 then the standard error of difference of two sample proportions is given by

$$\textit{Standard Error} = \frac{P_1Q_1}{n_1} + \frac{P_2Q_2}{n_2}$$

Important use of Standard Error

Standard error plays an important role in statistical theory. Following are its important uses.

- ☆ To measure the precision of a statistic. Higher the standard error lower its precision.
- ☆ To fix confidence limits for population parameters.
- ☆ To determine the size of the sample required to achieve the desired precision.
- ☆ To compute test statistic in tests of significance.

13. Central Limit Theorem

The central limit theorem, a fundamental statistical theory, states that when a large sample size is drawn from a population with a finite variance, the distribution of sample means will tend to be normally distributed. Furthermore, the mean of these sample means will be approximately equal to the mean of the entire population. Specifically, if a random sample of n observations is drawn from a population with mean 'm' and standard deviation 'σ', then the distribution of the sample mean '$\bar{x}$' approaches a normal distribution with mean 'm' and standard deviation '$\sigma/\sqrt{n}$' as 'n' increases.

It is important to note that the central limit theorem does not require the population from which the sample is drawn to be normally distributed. Even if the population is non-normal, the distribution of the sample mean will still tend towards normality as the sample size increases. This theorem holds a significant place in statistical inference, particularly when making inferences about populations based on large random samples.

Assumptions of Central Limit Theorem

- The sample should be drawn randomly following the condition of randomisation.
- The samples drawn should be independent of each other. They should not influence the other samples.
- When the sampling is done without replacement, the sample size shouldn't exceed 10% of the total population.
- The sample size should be sufficiently large.

Applications of Central Limit Theorem:

- The central limit theorem states that the sample distribution tends to be approximately normal, even when the population distribution is unknown or not normally distributed. This theorem does not require the population to be normally distributed. It is a fundamental concept in statistical analysis.
- As the sample size increases, the standard deviation of the sample mean decreases, leading to a more precise estimation of the population mean.
- The central limit theorem is frequently applied in election polls to estimate the proportion of people supporting a specific candidate and to construct confidence intervals around this estimate.
- The central limit theorem is also utilized in calculating various population parameters, such as the mean family income in a country, by providing reliable estimates and confidence intervals.
- By examining the sample distribution, the central limit theorem can help determine whether the sample is likely to be representative of the underlying population.
- The central limit theorem plays a crucial role in drawing conclusions about both sample and population parameters, aiding in the development of robust machine learning models and statistical analyses.

14. Statistical Inference

Statistical inference is a method of making decisions about the parameters of a population, based on random sampling. It helps to assess the relationship between the dependent and independent variables. The purpose of statistical inference is to estimate the uncertainty or sample to sample variation. The components used for making statistical inference are:

- ❑ Sample Size
- ❑ Variability in the sample
- ❑ Size of the observed differences

Statistical Inference Procedure

i. **Start with a Theory:** Begin by formulating a theory or hypothesis that explains the phenomenon being studied. This theory serves as the foundation for the entire procedure.

ii. **Design Research Hypothesis:** Based on the theory, formulate a specific research hypothesis that can be tested. This hypothesis is the central question being addressed by the study.

iii. **Operationalize the Variables:** Clearly define and operationalize the variables involved in the study. This includes identifying the dependent and independent variables and how they will be measured.

iv. **Recognize the Population:** Identify the population to which the study results should be applied. This is the set of all elements of interest in the study.

v. **Formulate a Null Hypothesis:** Create a null hypothesis that is the opposite of the research hypothesis. This null hypothesis is used to test the significance of the observed differences.

vi. **Assemble the Sample:** Draw a sample from the population. This sample is used to test the research hypothesis and draw conclusions about the population.

vii. **Conduct Statistical Tests:** Apply statistical tests to the sample data to examine if the collected sample properties are significantly different from what would be expected. These tests help determine whether the observed differences are due to chance or are statistically significant.

Importance of Statistical Inference

Inferential Statistics is crucial for thorough data examination and accurate conclusion-making. Proper data analysis is essential for interpreting research results effectively. It is predominantly utilized for future predictions across various fields and enables us to infer about the data. Statistical inference finds extensive applications in different fields

- ❑ Business Analysis
- ❑ Financial Analysis
- ❑ Machine Learning
- ❑ Pharmaceutical Sector
- ❑ Artificial Intelligence
- ❑ Social Sciences
- ❑ Share Market

Descriptive Statistics Vs Inferential Statistics

Category	Descriptive Statistics	Inferential Statistics
Purpose	Summarize and describe the main features of a dataset	Make inferences, predictions, or generalizations about a population based on sample data
Scope	Focuses on specific sample data	Extends findings to a larger population
Objective	Describe characteristics of the data without generalizing	Generalize findings from sample to population
Data Analysis	Provides a summary and visualization of data	Draws conclusions, tests hypotheses, and makes predictions
Statistical Techniques	Mean, median, mode, range, variance, standard deviation, histograms, box plots, etc.	Hypothesis testing, regression analysis, confidence intervals
Goal	To provide insights into the characteristics of a dataset	To make predictions or draw conclusions about a population
Examples	Measures of central tendency, dispersion, frequency distributions, graphical representations	Hypothesis testing, regression analysis, confidence intervals

15. General principles of testing of hypothesis

Statistical Hypothesis

- Statistical Hypothesis is a statement about the population under study.
- It is usually a statement about one or two parameters of the population.
- Such statements may or may not be true.

Example

- Feed A and B are equally effective in increasing the weight of fish.
- The probability of getting number 4 when a die is rolled - 1/6.

Null Hypothesis

The Null Hypothesis, often denoted as H_0, is a statement in statistical hypothesis testing that suggests there is no significant difference, effect, or relationship between the variables being studied.

Alternative Hypothesis

The Alternative Hypothesis, often denoted as H_1 or H_a, is a statement in statistical hypothesis testing that suggests there is a significant difference, effect, or relationship between the variables being studied. It contrasts with the null hypothesis, which proposes no significant difference or effect.

Test Statistic

It is a function of sample means. It extracts the information about the population parameter contained in the sample. The observed value of the test statistic serves as a guide in rejecting or not rejecting the null hypothesis. For example, in testing of null hypothesis that value of the population mean is m_0 i.e. H_0: $m = m_0$ the statistic used is

$$Z = \frac{(\overline{x} - m_o)}{\frac{\sigma}{\sqrt{n}}}$$

Rejection Region

After the test statistic to be used is selected, the set of possible values of a statistic is decided into two mutually exclusive regions viz: rejection region (critical region) and acceptance region (Region of non-rejection). If the observed value of the test statistic falls in the rejection region, H_0 is rejected. If it falls in the acceptance region, it is not rejected. It may be noted that if the observed value falls in the acceptance region, it does not prove the hypothesis, it simply fails to disprove it.

Degrees of Freedom

The number of independent observations available from the data for the estimation of a particular parameter or a quantity is called the degree of freedom.

It can be calculated by deducting from the number of observations and the number of constants that are calculated from the data. For instance, the estimate of population variance based on a sample of 'n' observations is given by

$$s^2 = \frac{\sum (x_i - \overline{x})^2}{n-1}$$

In this case, the constant (parameter), the population mean, is estimated by the sample mean $\overline{x}$. Hence, deduct 1 from the total number of observations 'n' to get the degrees of freedom, i.e., the degrees of freedom of s^2 will be n-1.

Types of Errors

Two kinds of errors are likely to be committed in testing a hypothesis.

- Type I error: If the null hypothesis is rejected when it is actually true, then such error is called type I error.
- Type II error: If a null hypothesis is accepted when it is false, then such error is called type II error.

Actual Situation	**Test Decision**	
	Accept H_0	**Reject H_0**
H_0 true	Correct decision probability = 1- α	Type I error probability = α
H_0 false	Type II error probability = β	Correct decision probability = 1 - β

16. Tests of Significance

I. Test of Hypothesis for Large Sample (n > 30)

If a sample of size n is draw from a normal population with mean 'm' and standard deviation 'σ' then the sample mean $\bar{x}$ is also distributed as normal with mean 'm' and standard deviation '$\frac{\sigma}{\sqrt{n}}$'.

As $\bar{x}$ is distributed with mean 'm' and standard deviation '$\frac{\sigma}{\sqrt{n}}$'. the standard normal variate is given by,

$$Z = \frac{(\bar{x} - m_0)}{\frac{\sigma}{\sqrt{n}}}$$

Under null hypothesis, H_0: m = m_0 then the test statistic,

$$Z = \frac{(\bar{x} - m)}{\frac{\sigma}{\sqrt{n}}}$$

(i) Critical Region or Rejection Region

The area under the normal curve between m ±1.96 σ is 0.95.

Hence, in the case of standard normal variable which has mean 0 and standard deviation 1, the area between -1.96 to +1.96 with the probability of 0.95.

In the test involving normal distribution, the set of values of Z outside the range -1.96 to +1.96 constitutes the region of rejection or critical region.

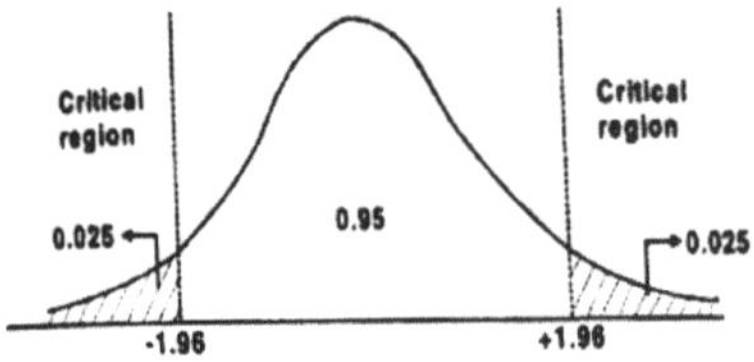

Critical Region, 5% level of significance

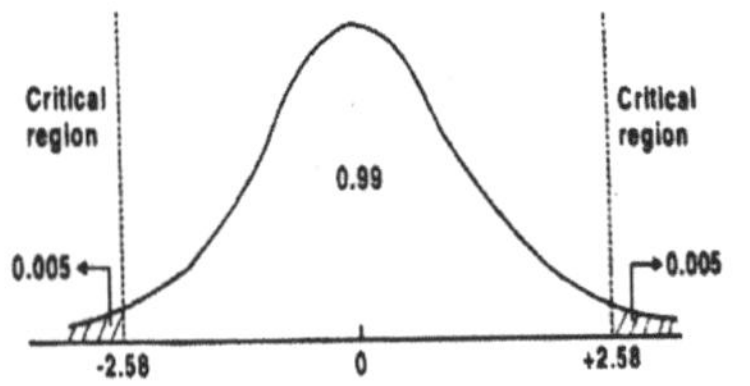

Critical Region, 1% level of significance

(ii) Two Tailed and One Tailed Test

If the null hypothesis $H_{0:}$ $m = m_0$ is tested against $H_{1:}$ $m = m_0$ (which implies $m > m_0$ or $m < m_0$), then the interest is on extreme values of Z on both tails of the distribution. In such cases the critical region is split between two sides or tails of the distribution of the test statistics. Test applied for such situation are called 'two tailed test'.

If the null hypothesis $H_{0:}$ $m = m_0$ is tested against $H_{1:}$ $m > m_0$ (which implies $m > m_0$ or $m < m_0$), then the interest is on extreme values to one side of the mean. In such cases the critical region will be to one side of the distribution. Test applied for such situation are called 'one tailed test'.

It is to be noted that critical value of Z at 5% and 1% level of significance for one tailed test are 1.645 and 2.33, whereas, these values are 1.96 and 2.58 for two tailed tests.

(iii) Test for Single Mean

Let $x_{1,}$ $x_{2,}$ x_3 ...... x_n be the values of a variable x, in a large random sample of size 'n' from a population with mean 'm' and variance 'σ^2'. On the basis of this sample, the hypothesis regarding the value m is tested. Testing of hypothesis consists of the following steps:

- **Hypothesis:** $H_{0:}$ $m = m_{0;}$ H_1: $m \neq m_0$, where m_0 is a specified value.
- **Test statistics:** $Z = \dfrac{(\overline{x} - m_0)}{\frac{\sigma}{\sqrt{n}}}$, where $\overline{x}$ is the sample mean.
- **Statistical decision:**
  - ☆ If $|z|>1.96$, reject the H_0 at 5% level of significance, otherwise do not reject.
  - ☆ If $|z|>2.58$, reject the H_0 at 1% level of significance, otherwise do not reject.

Example

A random sample of 200 fishes drawn from a certain species showed a mean length of 28 cm. Can this be considered as a sample from a population with mean 30 cm and standard deviation 14 cm?

Solution

- ❑ Hypothesis

  $H_0: m = 30;$

  $H_1: m \neq 30$

- ❑ Test statistic

Test statistics used is

$$Z = \frac{(\bar{x} - mo)}{\sigma}$$

$$= \frac{(31-32)\sqrt{200}}{14}$$

$$= \frac{-14.14}{14} = -1.01$$

- ❑ Statistical decision

Since $|z| < 1.96$, H_0 is not rejected at 5 % level of significance. Therefore, we do not have sufficient evidence to conclude that the sample comes from a population with a mean different from 32 cm.

(iv) Test for Equality of Two Population Mean

Let $\overline{x_1}$ be the mean of a large random sample size of n_1 from a population with mean m_1 and standard deviation σ^2 and let $\overline{x_2}$ be the mean of a large random sample size of n_2 from a population with mean m_2 and standard deviation σ^2. Testing of hypothesis consists of the following steps:

- ❑ **Hypothesis:** $H_{0:}\ m_1 = m_{2;}\ H_1: m_1\ m_2$
- ❑ **Test statistics:**

$$Z = \frac{\left(\overline{x_1} - \overline{x_2}\right)}{\sqrt{\frac{\sigma_1^{\ 2}}{n_1} + \frac{\sigma_2^{\ 2}}{n_2}}}$$

- ❑ **Statistical decision**
  - ☆ If $|z|>1.96$, reject the H_0 at 5% level of significance, otherwise do not reject.
  - ☆ If $|z|> 2.58$, reject the H_0 at 1% level of significance, otherwise do not reject.

Example

The mean length of 150 fish caught by sampling gear X was 10 cm with standard deviation of 1.5 cm, whereas the mean length of fish caught by sampling gear Y was 9.8 cm with standard deviation of 1.8 cm. Is there significance between the lengths of fish caught by the two gears at 5% level of significance.

Solution

- **Hypotheses**

If m_1 and m_2 denote the mean length of fish caught by gears X and Y respectively then

$H_0: m_1 = m_2$;

$H_1: m_1 \neq m_2$

- **Test statistic**

$$Z = \frac{\bar{x} - mo}{\sqrt{\frac{\sigma_1{}^2}{n_1} + \frac{\sigma_2{}^2}{n_2}}}$$

$$= \frac{10-9.8}{\sqrt{\frac{1.5^2}{150} + \frac{1.8^2}{150}}}$$

$$= \frac{0.2}{\sqrt{0.00015+0.000216}}$$

$$= \frac{0.2}{\sqrt{0.000366}}$$

$$= \frac{0.2}{0.01914}$$

$$Z = 10.446$$

- **Statistical decision**

Since $|z|>1.96$, H_0 is rejected at 5 % level of significance. Therefore, we have sufficient evidence to conclude that there is a significant difference between the lengths of fish caught by gears X and Y at the 5% level of significance

(v) Testing Single Population Proportion

Testing hypothesis about populations is carried out in the same way as for means when the conditions necessary for using normal distribution are met with. When a sample is sufficiently large,

- **Hypothesis:** $H_{0:}\ p = p_{0;}\ H_1: p \neq p_0$
- **Test statistics:**
- **Statistical Decision:** $Z = \frac{p - p_0}{\sqrt{\frac{p_o q_o}{n}}}$

☆ If $|z|>1.96$, reject the H_0 at 5% level of significance, otherwise do not reject.

☆ If $|z|>2.58$, reject the H_0 at 1% level of significance, otherwise do not reject.

Example

In survey of 800 households, 320 said they are fish eaters. Can it be concluded that in the sampled population the proportion who are fish eaters is 0.4?

Solution

- **Hypothesis**

R. block H_0 : $P = 0.4$;

$H_1 : P \neq 0.4$

From the given data sample proportion p = 320/800 = 0.4

$p_o = 0.4$,

$q_o = 0.6$

The test statistic used is

$$= \frac{(p - p_0)\sqrt{n}}{\sqrt{p_0 q_0}}$$

$$= \frac{(0.4 - 0.4)\sqrt{800}}{\sqrt{(0.4)(0.6)}}$$

= 0

As |z|<1.96 and also 2.58, the null hypotheses is not rejected at 5% and also 1% level of significance. Therefore, we do not have sufficient evidence to conclude that the proportion of fish owners in the sampled population is different from 0.4.

(vi) Test for Equality of two Population Proportion:

Let x_1 and x_2 be the number of individuals possessing the given attribute say A in random samples of size n_1 and n_2 from the two populations respectively. Sample proportions are given by,

$$p_1 = x_1 / n_1 \text{ and } p_2 = x_2 / n_2$$

If p_1 and p_2 are the proportions of populations 1 and 2 respectively then variance of $(p_{1-} p_2)$ is given by,

$$V(p_{1-} p_2) = \frac{P_1 + Q_1}{n_1} + \frac{P_2 + Q_2}{n_2} = PQ\left(\frac{1}{n_1} + \frac{1}{n_2}\right)$$

Under the null hypothesis $P_1 = P_2 = P$, say Q1= Q2 = Q.

- ☆ Hypothesis: $H_{o:}\ P_1 = P_2 = P$; H_1: $P_1 \neq P_2$
- ☆ Test statistics:

$$Z = \frac{P_1 - P_2}{\sqrt{\hat{P}\hat{Q}\left(\frac{1}{n_1} + \frac{1}{n_2}\right)}} \text{, Where } \hat{P} = \frac{n_1 p_1 + n_2 p_2}{n_1 + n_2} = \frac{x_1 + x_2}{n_1 + n_2}$$

Example

Two treatment X and Y were applied on a sample of 120 and 150 fish respectively for prevention of a particular disease. After a month it was observed that 20 and 30 fish which received X and Y treatments respectively were affected by disease. Test the hypotheses that both the treatments are equally effective in prevention of disease.

Solution

Let P_1 and P_2 denote population proportion of fish affected by disease using A and B treatments respectively.

- ❑ Hypotheses

$H_0: P_1 = P_2; \; H_1: P_1 \neq P_2$

- ❑ Test statistic

$$Z = \frac{p_1 - p_2}{\sqrt{\hat{P}\hat{Q}\left(\frac{1}{n_1} + \frac{1}{n_2}\right)}}$$

Given $x_1 = 20, x_2 = 30, n_1 = 120, n_2 = 150$

$$p_1 = \frac{x_1}{n_1} = \frac{20}{120} = 0.1667, \qquad p_2 = \frac{x_2}{n_2} = \frac{30}{150} = 0.2$$

$$\hat{P} = \frac{n_1 p_1 + n_2 p_2}{n_1 + n_2} = \frac{(120 \times 0.1667) + (150 \times 0.2)}{120 + 150} = 0.1852,$$

$$\hat{Q} = 1 - \hat{P} = 1 - 0.1852 = 0.8148$$

Hence,

$$Z = \frac{0.1667 - 0.2}{\sqrt{(0.1852)(0.8148)\left(\frac{1}{120} + \frac{1}{150}\right)}}$$

$$= \frac{-0.0333}{\sqrt{(0.1852)(0.8148)(0.0083 + 0.0067)}}$$

$$= \frac{-0.0333}{\sqrt{0.001249}}$$

$$= \frac{-0.0333}{0.03534}$$

$$= -0.9414$$

Statistical Decision

Since |Z|<1.96, we fail to reject the null hypothesis at the 5% level of significance. Therefore, we do not have sufficient evidence to conclude that both treatments are not equally effective in preventing the disease.

II. Test of Hypothesis of Small Samples (n<30)

When the size of the sample is small, the distribution of various statistics is far from normality and hence tests of hypothesis based on normal variate cannot be apply.

In such cases, tests of hypothesis based on exact sampling distribution of 't' and 'F' are applied.

(a) T distribution or Student 't' test

This distribution is symmetrical about the mean but is slightly flatter than the normal distribution. When the size of the sample is very small (<30), the t-distribution markedly differs from a normal distribution, but as n increases, t-distribution resembles more of normal distribution. The variable t-ranges theoretically from $-\infty$ to $+\infty$ The values of 't' have been tabulated for different degrees of freedom at different level of significance (Fisher and Yates, 1963).

(i) Test for Single Mean

Let $x_1, x_2, x_3 \ldots\ldots x_n$ be a random sample of size 'n' from a normal population with mean 'm'. Let and " denote mean and variance of sample. Testing of hypothesis consists of the following steps:

- Hypothesis: $H_0 : m = m_0;\ H_1 : m \neq m_0$
- Test statistics: $t = \frac{(\bar{x}-m_0)}{\frac{s}{\sqrt{n}}}$
- This t follows t – distribution with (n-1) degrees of freedom.
- Statistical decision:
  - ☆ If |t > the critical value of t at 5% level of significance, then reject the H_0 at 5% level of significance, otherwise do not reject.
  - ☆ If |t| > the critical value of t at 1% level of significance, then reject the H_0 at 1% level of significance, otherwise do not reject.

Example

A sample of 30 fingerlings drawn from a rearing tank showed a mean length of 78.5 mm and standard deviation of 12 mm. Is the data consistent with the claimed mean size of 82 mm?

Solution

Let m denote the population mean

- Hypotheses

$H_0 : m = 82;$

$H_1 : m \neq 82$

- **Test statistics:**

$$t = \frac{(\bar{x} - m_0)}{\frac{s}{\sqrt{n}}}$$

$$t = \frac{(78.5 - 82)}{\frac{12}{\sqrt{30}}}$$

$$t = \frac{-3.5}{\frac{12}{5.48}}$$

$$t = \frac{-3.5}{2.191} = -1.598$$

- **Statistical decision:**

Since, |t|<2.045, we fail to reject the null hypothesis at the 5% level of significance. Therefore, we do not have sufficient evidence to conclude that the mean size of the fingerlings is different from 82 mm.

(ii) Test of difference between two means of correlated observations (paired t-test)

When two samples of equal size are drawn from normal populations and these samples are not independent, then the paired t-test is used. Let 'n' be the size of each of the two samples and $d_1, d_2, \ldots\ldots d_n$ be the difference between the corresponding members of the sample. Let $\bar{d}$ denote the mean of differences and 's' be the standard deviation of these differences.

- **Hypothesis:**

  $H_0: m_1 = m_2;$

  $H_1: m_1 \neq m_2,$

  where m_1 and m_2 are means of 1st and 2nd population respectively.

- **Test statistics**

$$t = \frac{\bar{d}}{\frac{s}{\sqrt{n}}} \text{ Where, } \qquad s = \sqrt{\frac{1}{n-1}\left(\sum d^2 - \frac{(\sum d)^2}{n}\right)}$$

This t follows t – distribution with (n-1) degrees of freedom.

- **Statistical decision:**

If |t|> the critical value of t at a specified level of significance, then reject the H_0 at that level.

Example:

The following table gives the scores obtained by 12 students in two tests of Marine Biology, one conducted at the beginning of a semester and the other at the end of the semester after intensive coaching. Do the data indicate that the students have benefited from coaching?

Student	1	2	3	4	5	6	7	8	9	10	11	12
Test 1	72	80	65	75	79	85	60	70	85	90	78	82
Test 2	85	90	75	80	85	92	70	78	90	95	88	88

Solution

- **Hypothesis:**

  $H_0: m_1 = m_{2;}$

  $H_1: m_1 \neq m_2,$

where m_1 and m_2 are mean marks before and after training

- **Test statistics:**

Student	1	2	3	4	5	6	7	8	9	10	11	12	Total
Test 1	72	80	65	75	79	85	60	70	85	90	78	82	
Test 2	85	90	75	80	85	92	70	78	90	95	88	88	
Difference (di)	13	10	10	5	6	7	10	8	5	5	10	6	85
di2	169	100	100	25	36	49	100	64	25	25	100	36	839

$$\bar{d} = \frac{\sum d_i}{9} = \frac{85}{12} = 7.083$$

$$s = \sqrt{\frac{1}{n-1}\left(\sum d2 - \frac{\left(\sum d\right)2}{n}\right)}$$

$$= \sqrt{\frac{1}{12-1}\left(839 - \frac{85^2}{12}\right)}$$

$$= \sqrt{\frac{1}{11}\left(839 - \frac{7225}{12}\right)}$$

$$= \sqrt{\frac{236.92}{11}}$$

$$= \sqrt{21.538}$$

$$= 4.68$$

Therefore, $t = \frac{\bar{d}}{\frac{s}{\sqrt{n}}} = \frac{7.083}{\frac{4.64}{\sqrt{12}}} = \frac{7.083}{1.343} = 5.267$

- **Statistical decision:**

The table value of t with 11 degrees of freedom is 2.306 at 5% level of significance.

Since |t|>2.201, we reject the null hypothesis at the 5% level of significance. Therefore, the data indicates that the students have benefited from coaching.

(iii) Testing of Difference Between Two Means (Population Variance Assumed Equal)

Let $\bar{x}$ and s_1 be the mean and standard deviation of a sample of size n_1 from a normal population with mean m_1 and let $\bar{Y}$ and s_2 be the mean and standard deviation of a another sample of size n_2 from a normal population with mean $m_{2.}$ To test whether the population means differ significantly, the following null hypothesis is set up:

- Hypothesis

  H_0: $m_1 = m_{2,}$

  H_1: $m_1 = m_2$

- **Test statistics:**

$$t = \frac{\bar{X} - \bar{Y}}{s\sqrt{\frac{1}{n_1} + \frac{1}{n_2}}}$$

which is distributed at t with $n_1 + n_2 - 2$ degrees of freedom.

s in the above expression is computed using the formula

$$s = \sqrt{\frac{\sum(X-\bar{X})^2 + \sum(Y-\bar{Y})^2}{n_1 + n_2 - 2}}$$

$$= \sqrt{\frac{(n_1 - 1)s_1^2 + (n_2 - 1)s_2^2}{n_1 + n_2 - 2}}$$

- **Statistical decision:**

If > the critical value of t at a specified level of significance, then reject the H_0 at that level.

Example

Weight was recorded separately for male and female of one year old fishes. The mean weights of males and females are:

Sex	Sample Size	Mean weight (g)	Variance
Male	12	55	36
Female	15	50	25

Test whether there is significant difference in mean weight of male and female fish.

Solution

Let m_1 and m_2 denote the population means of male and female fishes respectively.

- **Hypothesis:**

  $H_0: m_1 = m_2,$

  $H_1: m_1 = m_2$

- **Test statistics:** $\dfrac{\bar{X}-\bar{Y}}{s\sqrt{\dfrac{1}{n_1}+\dfrac{1}{n_2}}}$

Given, $\bar{X}=55, \bar{Y}=50, s_1^{\,2}=36, s_2^{\,2}=25, n_1=12, n_2=15$

$$s = \sqrt{\frac{(n_1-1)s_1^{\,2}+(n_2-1)s_2^{\,2}}{n_1+n_2-2}}$$

$$s = \sqrt{\frac{(11)36+(14)25}{25}}$$

$$s = \sqrt{29.84}$$

$$s = 5.46$$

$$t = \frac{55-50}{5.46\sqrt{\frac{1}{12}+\frac{1}{15}}}$$

$$= \frac{5}{5.46\sqrt{\frac{27}{180}}}$$

$= 2.366$

- **Statistical decision:**

  ☆ If $|t| > t_{table}$ at a specified level of significance, then reject H_o at that level.

  ☆ At a 5% level of significance (two-tailed test) and with df=12+15−2=25, t_{table} = 2.064.

Since |2.366|>2.064, we reject H_o at the 5% level of significance. Therefore, there is a significant difference in the mean weight of male and female fish.

(b) Chi-Square Distribution

The X^2 distribution can be defined as the sum of squares of independent normal variates. If $X_1, X_2, \ldots\ldots X_n$ are n independent standard normal variates, then sum of squares of these variates $X_1^{\,2} + X_2^{\,2} + \ldots\ldots + X_n^{\,2}$ follows the X^2 distribution with n degrees of freedom.

The shape of X^2 distribution depends on the degrees of freedom which is also its mean. When 'n' is small, the X^2 distribution is markedly different from normal distribution but as 'n' increases the shape of the curve becomes more and more symmetrical and for n>30, it can be approximated by a normal distribution.

The values of X^2 have been tabulated for different degrees of freedom at different levels of probability (Fisher and Yates, 1963). The X^2 always greater than or equal to zero ($X^2 \geq 0$). The X^2 test is commonly used for analysis of attribute data.

Test for Fixed-ratio Analysis

The common carps, normal pigmentation is due to a dominant allele BB and its recessive allele bb produces blue pigmentation. F_1 generation will have all the individual (Bb) with normal pigmentation. When F_1 are crossed with F_2 generation there will be common carps with normal and blue pigmentation in the ration of 3:1.

Whether this hypothesis of 3:1 ratio is substantiated by the actual observed data can be ascertained by X^2 test. This X^2 test can be applied to test any fixed ratio hypothesis provided the expected ratio is specified before the investigation commences.

If O_i refers to observed frequency and E_i refers to the expected frequency based on the expected ratio hypothesis, X^2 is computed as follows:

$$X^2 = \sum_{i=1}^{k} \frac{(O_i - Ei)^2}{E_i} \Longrightarrow X^2 = \sum_{i=1}^{k} \frac{O_i^2}{E_i} - n$$

When 'n' is the total number of observation and k is the number of classes. The X^2 has k-1 degrees of freedom. In this test the expected frequency of each class should be more than 5. If any such frequency is small adjacent classes may be grouped, so that the expected frequency is more than 5.

If the calculated value of X^2 is greater than the table value of X^2 with (k-1) degrees of freedom, at specified level of significance the null hypothesis of specified ratio is rejected.

Goodness of Fit Test for Probability Distributions

An important application of the chi-square test is to test whether a set of quantitative data follows a specific probability distribution. In this test, the actual frequencies observed in each category are compared with the frequencies that would be theoretically expected if the data followed the hypothesized probability distribution. To perform this test, the following steps are followed:

- Hypothesize the probability distribution to be fitted.
- Estimate the values of each parameter of the selected probability distribution from the given data, if not specified.
- Estimate the theoretical frequencies for each category based on the hypothesized probability distribution.
- Compute the chi-square test statistic using the formula:

$$X^2 = \sum_{i=1}^{k} \frac{O_i^2}{Ei} - n$$

☆ If the expected frequency of any class is less than 5, the adjacent classes can be grouped to form a class, so that expected frequency is more than 5.

☆ If the expected frequencies are calculated on the basis of certain parameters estimated from data, the degrees of freedom for X^2 is not (k-1) but is decreased by the number of parameters estimated.

☆ If X^2 computed in step. 4 is greater than the tabular value of X^2 with (k-1) degrees of freedom at specified level of significance, the null hypothesis for the given data is rejected.

Example

A sample of 700 fish observed for determining the sex ratio, indicated that 380 were male and 320 were female. Do the observed data fit the expected ratio 3:4?

Solution

❑ **Hypothesis:**

H_0: The observed data fit the ratio 3:4

H_1: The observed data does not fit the ratio 3:4

❑ **Test statistic:**

On the basis of this hypothesis of 3:4 ratio, 300 fishes are expected to be male and 400 fishes are expected to be female classes. X^2 is calculated as follows:

Sex	Frequency		O_i^2	$\frac{O_i^2}{E_i}$
	Observed (O_i)	Expected (E_i)		
Male	380	300	1,44,400	481.33
Female	320	400	1,02,400	256
Total	700	700		737.33

$$X^2 = \sum \frac{O_i^2}{E_i} - n$$

$$= 737.33 - 700$$

$$= 37.33$$

❑ **Statistical decision:**

The table value of X^2 with 1 df at 5% level of significance is 3.841. Since the computed X^2 value is greater than the table value of X^2, we reject the null hypothesis. Therefore, the observed data does not fit the expected ratio of 3:4.

(i) X^2 Test for Independence of Attributes in 2x2 Contingency Table:

Suppose that an attribute data of size n is classified according to two attributes, say A and B and the attribute A is further subdivided into two classes A1 and A2 and the attribute B into B1, and B2. Such attribute data can be presented in the form of a table called 2x2 contingency table.

B/A	A1	A2	Total
B1	a	b	a+b
B2	c	d	c+d
Total	a+c	b+d	a+b+c+d = n

It may be of interest to test the independence of attributes A and B.

H_0: The two attributes A and B are independent

H_1: The two attributes A and B are dependent

A formula for computing X^2 of 2x2 contingency table is given by,

$$X^2 = \frac{n(ad-bc)^2}{(a+b)(c+d)(a+c)(b+d)}$$

- Where a,b,c and d are cell frequencies of 2x2 contingency table and n is the total frequency. It has 1 degree of freedom.
- If the expected cell frequencies are large, the discrete distribution of probabilities of all frequencies approximates the normal distribution. This approximation holds fairly well when the degrees of freedom are more than 1 and the expected frequency in various classes is not small.

As the degrees of freedom of X^2 statistic of 2x2 contingency table is 1, X^2 approximation in this case will not be satisfactory and leads to over estimation of significance. This is corrected by the method Yates suggested, known as 'Yates Correction'. The correction consists of adding ½ to the observed minimum frequency and adjusting the other cell frequency for the observed marginal totals and then computing the X^2.

Formula for X^2 using the Yates correction in a 2x2 contingency table is given by,

$$X^2 = \frac{n[|ad-bc|-\frac{n}{2}]^2}{(a+b)(c+d)(a+c)(b+d)}$$

This correction is suitable when the expected frequency of classes is less than 5, but estimation with correction can do no harm even when the frequencies are large. Hence, it is always better to use the correction as a matter of routine.

Example

In a series of experiments to test whether a new medication is effective in curing a particular infection, the following observations were found:

	Not cured	Cured	Total
Lime treated	72	28	100
Untreated (Control)	74	26	100
Total	146	54	200

Test whether the new medication has any effect in curing the infection.

Solution

- **Hypothesis:**

  H_0: There is no association between the medication and curing the infection.

  H_1: There is an association between the medication and curing the infection.

- **Test statistic:**

$$X^2 = \frac{n(ad-bc)^2}{(a+b)(c+d)(a+c)(b+d)}$$

Where a = 72, b = 28, c = 74, d = 26

$$X^2 = \frac{200(72\times 26-28\times 74)^2}{(100)(100)(146)(54)}$$

$$X^2 = \frac{200(-200)^2}{(100)(100)(146)(54)}$$

$X^2 = 0.101$

X^2 with Yates correction

$$X^2 = \frac{n\left[|ad-bc|-\frac{n}{2}\right]^2}{(a+b)(c+d)(a+c)(b+d)}$$

$$X^2 = \frac{200\left[|72\times 26-28\times 74|-\frac{200}{2}\right]^2}{(100)(100)(146)(54)}$$

$$X^2 = 0.0253$$

- **Statistical decision**

The table value of with 1 degree of freedom at a 5% level of significance is approximately 3.841. Since both the computed values are less than the table value of , we fail to reject the null hypothesis. Therefore, there is no evidence to suggest an association between the medication and curing the infection.

(c) Test of Hypothesis about a Population Variance

Let $x_1, x_2, ...x_n$ be the values of a variable in a random sample of size n drawn from a normal population with variance .

The null hypothesis to be tested is, H_0: $\sigma^2 = ,\sigma_0^2$ where σ_0^2 is a specified value. Test statistic used is,

$$x^2 = \frac{(n-1)\,s^2}{\sigma^2}$$

Where s^2 is the sample variance, X^2 in above equation is distributed as x^2 with (n-1) df.

If alternative hypothesis to be tested is

$H_1:\sigma^2\neq\sigma_0^2$, then reject H_0 if, $x^2_{Cal} > x^2_{\alpha/2}$ or $x^2 < x^2_{1-\alpha/2}$

$H_1: \sigma^2 > \sigma_2^2$, then reject H_0 if, $x^2_{Cal} > x^2_{\alpha}$

$H_1: \sigma^2 < \sigma_0^2$ then reject H_0 if, $x^2_{Cal} < x^2 1\text{-}\alpha$

Chapter-5

Data Analysis Techniques & Application in Fisheries Science

17. Bivariate Data

Bivariate analysis is a statistical method examining how two different things are related. It aims to determine if there is a statistical link between the two variables and, if so, how strong and in which direction the link is. It is a helpful technique for determining how two variables are connected and finding trends and patterns in the data. Typically, X and Y represent the two variables being analyzed. Bivariate data is represented as a pair, typically denoted as (X, Y).

Characteristics of Bivariate Data

i. **Two Variables**: Bivariate data involves exactly two variables, which can be either numerical or categorical.

ii. **Pairwise Relationship**: Each value of one variable is paired with a value of the other variable, creating a relationship between the two variables.

iii. **Types of Variables**: The variables can be of different types, such as:

- **Numerical and Numerical:** Both variables are numerical. For example, fishing gear size (in meters) and catch weight (in kilograms) are two numerical variables that can be paired to analyze the relationship between them.
- **Categorical and Categorical:** Both variables are categorical. For example, fishing gear type (e.g., trawl, long line, or gillnet) and fishing location (e.g., coastal, offshore, or estuary) are two categorical variables that can be paired to analyze the relationship between them.
- **Numerical and Categorical:** One variable is numerical and the other is categorical. For example, fishing experience (in years) and fishing gear type (e.g., trawl, longline, or gillnet) are a numerical and categorical

variable pair that can be analyzed to understand how experience affects the choice of fishing gear.

Analysis of Bivariate Data

Many kinds of bivariate analysis can be used to determine how two variables are related. Here are some of the most common types.

i. **Scatterplots:** A scatterplot is used to visualize the relationship between the two variables, where each point represents a pair of values from the two variables.

ii. **Correlation Coefficients:** Correlation coefficients are used to quantify the strength and direction of the relationship between the two variables. The correlation coefficient ranges from -1 (perfect inverse correlation) to 1 (perfect direct correlation), with 0 indicating no correlation.

iii. **Simple Linear Regression:** Simple linear regression is used to model the relationship between the two variables, providing a mathematical equation that describes how one variable changes in response to changes in the other variable.

Importance of Bivariate Data in Fisheries

i. **Identifying Trends and Patterns:** Analyzing the relationship between fishing effort (e.g., number of fishing vessels) and catch per unit effort (CPUE) can reveal trends in fish populations over time. For example, a declining CPUE despite stable effort may indicate overfishing.

ii. **Identifying Cause and Effect Relationships:** Studying the link between environmental variables like water temperature and fish abundance can help identify the causes of changes in fish populations. For instance, a correlation between rising ocean temperatures and declining fish stocks may suggest climate change is impacting fisheries.

iii. **Making Predictions:** Modeling the relationship between fishing effort and catch can enable predictions of future yields under different management scenarios. This information is crucial for setting sustainable catch limits and quotas.

iv. **Informing Decision-Making:** Bivariate analysis of fisheries data can guide policy decisions, such as implementing gear restrictions, closed seasons, or marine protected areas to rebuild depleted stocks. For example, a correlation between juvenile bycatch and a particular gear type may prompt a ban on that gear.

v. **Hypothesis Testing:** Researchers can use bivariate statistics to test hypotheses about fisheries, such as whether a new management measure has increased fish biomass. For instance, comparing CPUE before and after implementing a no-take zone can assess its effectiveness.

vi. **Optimizing Resource Allocation:** Bivariate analysis can help optimize the allocation of limited management resources. For instance, identifying the gear types or fishing locations with the highest bycatch rates can guide targeted enforcement efforts.

vii. **Evaluating Interventions:** Comparing fisheries data before and after an intervention, such as stocking a lake with fingerlings, can assess the impact of the management action. Bivariate analysis of stocking rates and subsequent catch per unit effort can quantify the effectiveness of stocking.

viii. **Communicating Findings:** Visualizations like scatterplots and correlation coefficients provide a clear way to communicate bivariate relationships in fisheries to stakeholders, such as the correlation between fishing effort and catch. These graphics can effectively convey complex information to inform management decisions.

18. Scatter plots

Scatterplots are a type of data visualization used to show the relationship between two numerical variables. Each data point is plotted as a dot, with its position determined by the values of the two variables. They display the values of one variable on the x-axis and the values of the other variable on the y-axis. The pattern depicted in a scatter plot indicates the type and strength of the relationship between the two variables. In a scatter plot, if the variables are correlated, the points tend to cluster along a curve or line. It's important to note that correlation does not imply causation; the relationship between variables may be influenced by other factors. A scatter diagram or scatter plot gives an idea of the nature of relationship.

1. Perfect Positive Correlation: A perfect positive correlation is a situation where the relationship between two variables is linear and the points on the scatterplot form a straight line with a positive slope. This means that as one variable increases, the other variable also increases. The correlation coefficient in such cases is 1, indicating a perfect positive correlation.

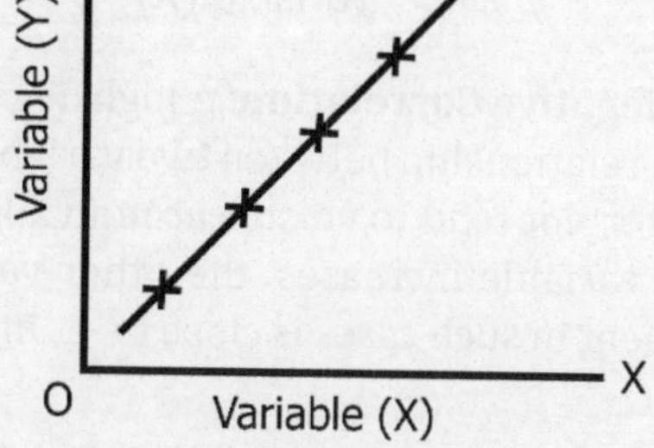

2. Perfect Negative Correlation: A perfect negative correlation is a situation where the relationship between two variables is linear and the points on the scatterplot form a straight line with a negative slope. This means that as one variable increases, the other variable decreases. The correlation coefficient in such cases is -1, indicating a perfect negative correlation.

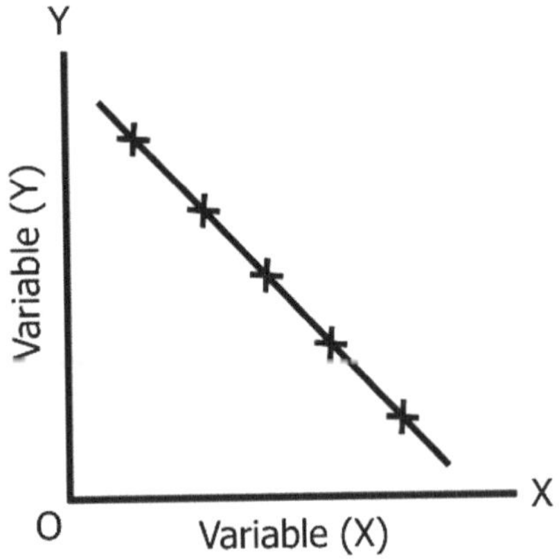

3. High Degree of Positive Correlation: A high degree of positive correlation is a situation where the relationship between two variables is strong and positive. The points on the scatterplot tend to cluster around a line with a positive slope, indicating that as one variable increases, the other variable also tends to increase. The correlation coefficient in such cases is close to 1, indicating a strong positive correlation.

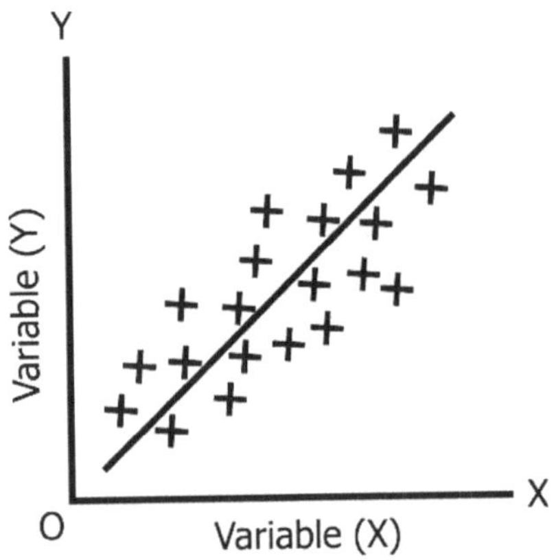

4. High Degree of Negative Correlation: A high degree of negative correlation is a situation where the relationship between two variables is strong and negative. The points on the scatterplot tend to cluster around a line with a negative slope, indicating that as one variable increases, the other variable tends to decrease. The correlation coefficient in such cases is close to -1, indicating a strong negative correlation.

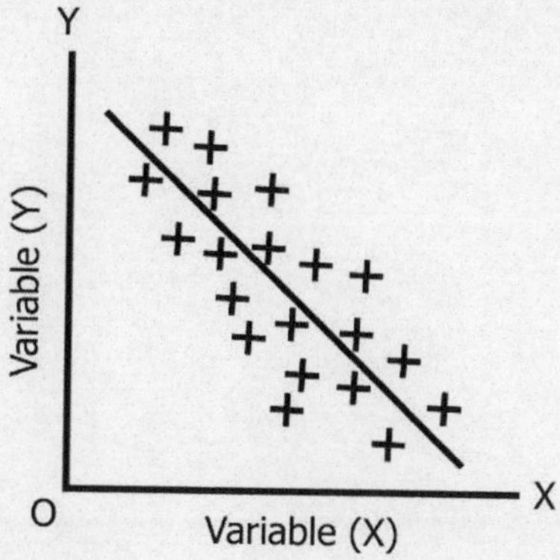

5. Direct Curvilinear: A direct curvilinear relationship is a situation where the points on the scatterplot form a curve that opens upwards. This means that as one variable increases, the other variable also increases, but the relationship is not linear. The correlation coefficient in such cases is positive but not equal to 1, indicating a curvilinear relationship.

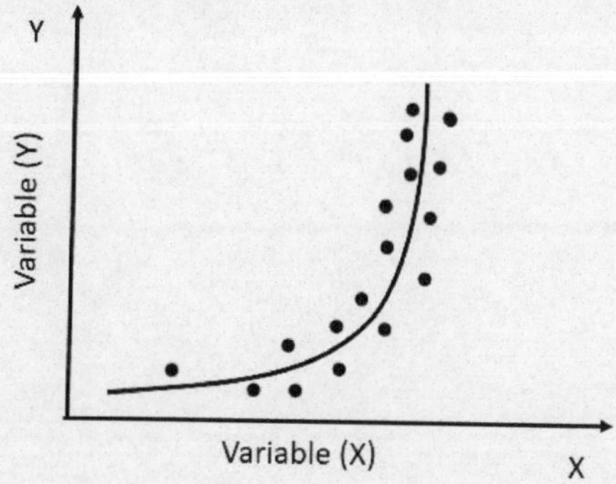

6. Inverse Curvilinear: An inverse curvilinear relationship is a situation where the points on the scatterplot form a curve that opens downwards. This means that as one variable increases, the other variable decreases, but the relationship is not linear. The correlation coefficient in such cases is negative but not equal to -1, indicating a curvilinear relationship

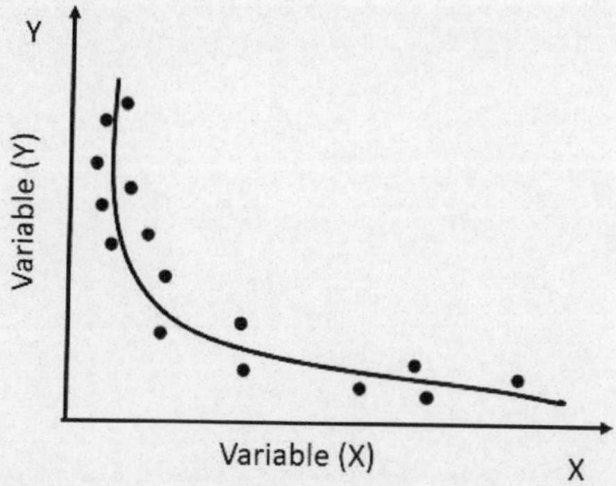

19. Simple Linear Correlation

It is a statistical tool to study the degree of association or relationship existing between two variables, when the relationship is linear or approximately linear.

The degree of relationship is quantified by a coefficient called the 'Karl Pearson Correlation Coefficient' or 'correlation coefficient'.

It's denoted by 'r'

$$r = \frac{\sum XY - \frac{(\sum X)(\sum Y)}{n}}{\sqrt{(\sum X^2 - \frac{(\sum X^2)}{n})(\sum Y^2 - \frac{(\sum Y^2)}{n})}}$$

In the above expression, X and Y denote the measurements on variables X and Y, n is the number of pairs of observation i.e. the sample size.

Types of Correlation:

Based on the direction of change in the value of two variables:

Positive Correlation:

When two variables move in the same direction; i.e., when one increases the other also increases and vice-versa, then such a relation is called a Positive Correlation.

For example, Relationship between the price and supply, income and expenditure, height and weight, etc.

Negative Correlation

When two variables move in opposite directions; i.e., when one increases the other decreases, and vice-versa, then such a relation is called a Negative Correlation.

For example, the relationship between the price and demand, etc.

Based on the Ratio of Variations Between the Variables

Linear Correlation

When there is a constant change in the amount of one variable due to a change in another variable, it is known as a Linear Relationship. This term is used when two variables change in the same ratio. If two variables that change in a fixed proportion are plotted on a graph, a straight line passing through the origin will represent the relationship between them. As a result, it suggests a linear relationship.

X	10	15	20	25	25
Y	10	20	30	40	50

In the above table, for every change of 5 units in variable X, there is a change of 10 units in variable Y. The ratio of change between variables X and Y in the above schedule is 1:2, and it remains constant. Thus, there is a linear relationship between the variables

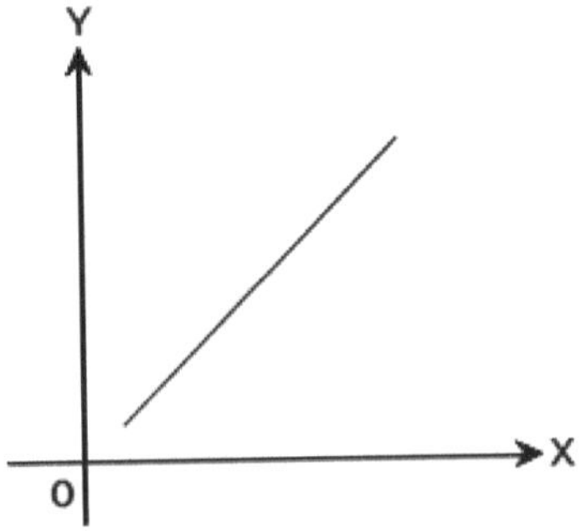

Linear Correlation

Non-Linear (Curvilinear) Correlation

When there is no constant change in the amount of one variable due to a change in another variable, it is known as a Non-Linear Relationship. This term is used when two variables do not change in the same ratio. This indicates that they do not form a straight-line relationship. For example, the production of grains would not necessarily increase even if the use of fertilizers is doubled.

X (Fertilizer)	10	20	30	40	50
Y (Production of Grains)	7	12	19	25	35

In the above example, there is no specific relationship between the variables. Although both variables change in the same direction (i.e., both are increasing), they change in different proportions. The ratio of change between variables X and Y in the above schedule is not constant, indicating a non-linear relationship between the variables.

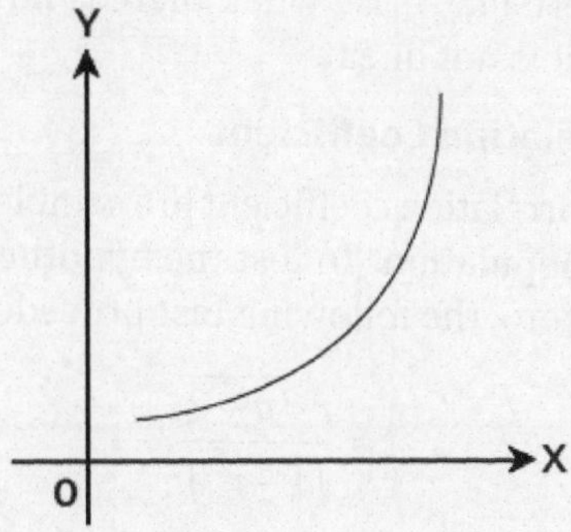

Non-Linear Correlation

Based on the Number of Variables Involved

Simple Correlation

Simple correlation implies the study between the two variables only. *For example,* the relationship between price and demand, and the relationship between price and money supply.

Partial Correlation:

Partial correlation implies the study between the two variables keeping other variables constant. *For example,* the production of fish depends upon various factors like water quality, quality of seeds, feed, etc. But, if one studies the relationship between fish and the quality of seeds, keeping water quality and feed constant, then it is a partial correlation.

Multiple Correlation:

Multiple correlation implies the study between three or more three variables simultaneously. The entire set of independent and dependent variables is studied simultaneously.

For example, the relationship between fish production with the quality of seeds and water quality.

Properties of Correlation Coefficient

- It is a pure number without units or dimensions
- It lies between -1 and +1, *i.e.* $-1 \leq r \leq +1$
- The correlation coefficient is independent of the origin and the scale of measurement of variables.
- The variables are said to be positively correlated if r is positive and negatively correlated if r is negative.
- When r = +1, there exists a strict linear relationship and the correlation between the variables is said to be perfectly positive.

- When r = - 1, the relationship is linear and correlation between the variables is perfectly negative.
- Value of r near zero may arise when there is no relationship or there is a relationship but it is not linear.

Significance of the Correlation Coefficient

Let r be the observed correlation coefficient in a sample of n pairs of observations from a bivariate normal population. To test the hypothesis $H_0 : \rho = 0$, that means population coefficient is zero, the following test procedure is used:

$$t = \frac{r\sqrt{n-2}}{\left(1-r^2\right)}$$

- Which is distributed as t with n-2 degrees of freedom. If the calculated value of t is greater than the table value of t with n-2 degrees of freedom, at the desired level of significance, the correlation between the variables is significant.
- However, it is to be noted that significance of r is not an indication of strength of relationship. It is simply a test to see whether ρ is equal to zero or not.
- The degree of the relationship between two variables can be measured by square of the correlation coefficient r^2 (which is called coefficient of determination).

Example-1

The total length and standard lengths of 20 fishes of a particular species were measured. Work out the coefficient of correlation for the given data.

S.No	**1**	**2**	**3**	**4**	**5**	**6**	**7**	**8**	**9**	**10**
X (Total Length)	112	105	118	122	148	120	128	135	168	138
Y (Standard Length)	85	80	90	94	115	92	100	110	135	105

S.No	**11**	**12**	**13**	**14**	**15**	**16**	**17**	**18**	**19**	**20**
X (Total Length)	148	176	158	172	165	145	132	142	155	150
Y (Standard Length)	115	140	125	130	127	120	102	110	121	118

Solution

S. No.	X	Y	XY	X^2	X^2
1	112	85	9520	12544	7225
2	105	80	8400	11025	6400
3	118	90	10620	13924	8100
4	122	94	11468	14884	8836
5	148	115	17020	21904	13225
6	120	92	11040	14400	8464
7	128	100	12800	16384	10000
8	135	110	14850	18225	12100
9	168	135	22680	28224	18225
10	138	105	14490	19044	11025
11	148	115	17020	21904	13225
12	176	140	24640	30976	19600
13	158	125	19750	24964	15625
14	172	130	22360	29584	16900
15	165	127	20955	27225	16129
16	145	120	17400	21025	14400
17	132	102	13464	17424	10404
18	142	110	15620	20164	12100
19	155	121	18755	24025	14641
20	150	118	17700	22500	13924
	2837	2214	320552	410349	250548

$$r = \frac{\sum XY - \frac{(\sum X)(\sum Y)}{n}}{\sqrt{\left(\sum X2 - \frac{(\sum X2)}{n}\right)\left(\sum Y2 - \frac{(\sum Y2)}{n}\right)}}$$

$$r = \frac{20 \times 320552 - (2837)(2214)}{\sqrt{\left[20 \times 410349 - (2837)^2\right]\left[20 \times 250548 - (2214)^2\right]}}$$

$$r = 0.9879$$

Example-2

Suppose the correlation coefficient is 0.2 and the number of observations is 32. What is the calculated test statistic? Is this significant correlation using a 5% level of significance?

Solution

- ❑ **Hypothesis**

  Ho: P = 0 there is no correlation

  H: P ≠ 0 there is correlation

- **Test statistics**

$$t = \frac{r\sqrt{N-2}}{\sqrt{1-r^2}}$$

Given, r = 0.2, N = 32

$$t = \frac{0.2\sqrt{32-2}}{\sqrt{1-(0.2)^2}}$$

$$t = 1.11803$$

- **Statistical decision**

Degree of freedom is N^{-2} = 32-2 = 30.

Since $t_{cal} < t_{tab}$, which means 1.11803 < 2.042. So that the null hypothesis is not rejected. Therefore, there is no significant correlation between the variables at the 5% level of significance.

20. Linear Regression

If two variables are found to be highly correlated, then a more useful approach would be to study the nature of their relationship. Linear regression aims to demonstrate the nature of the relationship between two variables by applying a best-fit linear equation to observed data, whereas the correlation coefficient estimates the degree or intensity of the relationship. One variable is considered the independent variable, and the other is considered the dependent variable. For example, a person's weight is often linearly related to their height, indicating a linear relationship between the two. As the height generally increases, the weight of the person also tends to increase. Simple linear regression focuses on studying linear relationships involving two variables, while relationships among more than two variables are studied using multiple regression techniques.

A Linear Regression Line Equation

A linear regression line equation is written in the form of:

$$Y = a + bX$$

Where, X is the independent variable and plotted along the x-axis

Y is the dependent variable and plotted along the y-axis

The slope of the line is b (regression coefficient), and a (Y intercept) is the intercept (the value of y when x = 0).

Two Regression Lines

If two variables X and Y are open to choices as to which affects which then 2 regression lines may be conceives.

Regression Equation of Y on X:

- If Y is considered as dependent variable, then the regression equation of Y on X is given by Y = a + bX
- The regression coefficient b is called the regression coefficient of Y on X (bYX).

Regression Equation of X on Y:

- If X is considered as the dependent variable, then the regression equation of Y on X is given by X = a + bY
- The regression coefficient b is called the regression coefficient of X on Y (bXY).

Properties of Regression Lines

- The regression lines intersect at point $(\bar{X}, \bar{Y})$
- If the variables are perfectly correlated, the regression lines coincide.
- If the variables are not correlated the regression lines are perpendicular to each other.

Applications of Linear Regression Analysis

- Conversion between different length measurements *i.e.*, from say, total length to standard length or to fork length and so on.
- Calculation of fish lengths from scale measurements.
- Fitting weight-length relationship
- Estimating parameters of the Von-Bertlanffy growth curve
- Splitting of total mortality coefficient Z into M, the natural mortality coefficient and F, the fishing mortality coefficient, using the regression of Z on effort f.
- Catch curve method of estimating instantaneous total mortality rate making use of the relation between abundance (number) of fish and their corresponding age.
- Schaefer's method of population analysis is based on the regression of catch per unit of effort on effort.
- Relation of fecundity to body weight using the regression of number of eggs (F) on body weight (W).
- Routine metabolic rate of fishes through the regression of oxygen consumption quiescent of fishes (Q) on body weight W.
- Estimation of heritability of different traits based on the regression of parents on offspring.

Relation Between Correlation and Regression

	Correlation	Regression
Meaning	Correlation is a statistical measure determining a co-relationship or association between two variables.	Regression describes how an independent variable (x) is numerically related to a dependent variable (y).
Main Purpose	Correlation analysis lets experimenters know the association or the absence of the relationship between two variables under study; if the variables are correlated, it allows measuring the strength of their association.	Regression analysis helps determine a functional relationship between two variable so as to estimate the unknown variable with the help of known variable(s) and make future projections on events.
Objective	To find a numerical value that expresses the relationship between the variables.	To estimate the values of a random variable on the basis of the value of a fixed variable.
Usage	Represents the linear relationship between two variables.	Fits the best line and estimates one variable on the basis of another variable.
Nature of variables	The variables are not designated as dependent or independent.	One variable is dependent, and another variable is independent.
Indicates	Correlation coefficient indicates the extent to which two variables move together.	Regression indicates the impact of a unit change in the known variable on the estimated variable.

Example

The data on fish yield tested under 5 stocking densities are given below:

Sr. No.	1	2	3	4	5
Fingerlings (1000/ha)	2	3	4	5	6
Fish yield (t/ha)	2.5	3.6	4.4	5.0	5.4

- ❑ Provide the regression equation
- ❑ Estimate fish yield for stocking density of 4200/ha.
- ❑ Calculate S^2.
- ❑ Perform test statistics

Solution

S. No	Fingerlings (x)	Fish yield (y)	xy	x^2	y^2
1	2	2.5	5	4	6.25
2	3	3.6	10.8	9	12.96
3	4	4.4	17.6	16	19.36
4	5	5.0	25	25	25
5	6	5.4	32.4	36	29.16
Total	20	20.9	90.8	90	92.83

(i) $b = n\Sigma xy - \Sigma x \Sigma y / n\Sigma x^2 - (\Sigma x)^2$

$= 5(90.8) - (20)(20.9)/5(90)^2 - (20)^2$

$= 454 - 418/450 - 400$

$b = 0.72$

$$\bar{y} = \frac{\sum y}{n} = 20.9/5 = 4.18; \bar{x} = \frac{\sum x}{n} = 20/5 = 4$$

$$a = \bar{y} - b\bar{x}$$

$= 4.18 - (0.7)(4)$

$a = 1.3$

Regression equation, $y = a + bx$

$y = 1.3 + 0.72x$

(ii) Estimate fish yield for stocking density of 4200/ha

x $= 4200/1000 = 4.2$

y $= 1.3 + bx$

$= 1.3 + (0.72).(4.2)$

$= 4.32$ tonnes

$S^2 = (n\text{-}1/n\text{-}2)\ Sy^2 - b^2 - Sx^2$

where, $Sy^2 = n\Sigma y^2 - (\Sigma y)^2/n(n\text{-}1) = 2.5$; $Sx^2 = 2.5$

$S^2 = 4/3\ (1.367 - (0.72)^2(2.5))$

$= (1.33).(0.071)$

$= 0.0947$

$H_0: b = 0$

$H_1: b \neq 0$

The test statistic is, $t = b/\sqrt{Sb^2}$

Where $Sb^2 = 1/n\text{-}2\ [(Sy^2/Sx) - b^2]$

$= 0.0095$

$t = 0.72\sqrt{0.0095} = 7.3846$

Table value at 5% C.L. with 3 d.f. is 3.18. Which is greater than calculated t value. So, H_0 is rejected.

21. Length Weight Relationship

The relationship between body weight (W) and body length (L) in fishes has been empirically observed be of the form

$$W=aL^b$$

This equation is not in linear form. The parameters 'a' and 'b' are almost universally estimated by researchers by transforming the equation to logarithmic form and applying the least square techniques. Thus the equation actually used is,

Log W=log a+b log L

The above mentioned tacitly assumes the following multiplicative error mode:

$W=aL^b{}_{e'}$

Where a and b are constants and e' is a random error factor

Taking logarithm to the base 'e' on both sides gives rise to

$\ln W = \ln a + b \ln L + \ln e'$

i.e $Y = A + BX + E$

Where

$Y = \ln W, X = \ln L, A = \ln a, b = B, E = \ln e'$

Expression in above equation is in the linear form. If it is assumed that E is distributed normally with mean zero and variance σ^2, then the estimates of A and B can be obtained by the method of least squares, using the formula

$$B=\frac{n\sum XY-\left(\sum X\right)\left(\sum Y\right)}{n\sum X^2-\left(\sum X\right)^2} \qquad A=\overline{Y}-B\,\overline{X}$$

In the above expressions Y = ln W and X = ln L, $\overline{Y}$ and $\overline{X}$ denote means of Y and X values respectively. The B value gives an estimate of b, whereas conventionally, 'a' is estimated as exp (A). This method however gives biased estimate of 'a'. To compensate for the bias the 'a' value is multiplied by the correction factor exp ($S^2/2$), where S^2 is an estimate of variance of deviations from regression. Hence, corrected a = exp ($A+S^2/2$)

$$a=\text{Antilog}\,(A+S^2/2)$$

Application of Length-Weight Relationship

- **Estimating Weight of Fish for a Given Length:** Length of fish can be more easily and accurately measured than weight in landing centers as well as on board the vessels in the sea. Hence, it is convenient to estimate weight from predetermined length-weight relationships.
- **Determining condition factor:** In order to compare weight and length in a particular sample or individual, condition factors are employed. Fulton's condition factor (K) is calculated as,

$$K = \frac{W}{L^3}$$

Where W and L are the observed total weight and length of a fish. It is the value of 'a' in length-weight relationship, $W = aL^b$, when b = 3. If the fish is heavier, at a given length, the larger is the factor K, implying better condition of fish. K is greater than 1, indicates general wellbeing of the fish is good. Fulton's condition factor, is suitable for comparing differences related to sex, season or place of capture.

Even when b differs from 3, Fulton's condition factor may be used, if fish are approximately of the same length. If the length range is large, the following formula is used:

$$K^* = \frac{W}{\widehat{W}}$$

Alternatively, the condition factor is computed as the ratio of observed weight to estimated weight.

$$K^* = \frac{W}{L^b}$$

Where is the estimated weight based on length-weight relationship.

$$W=aL^b$$

Example:

Total length (cm) and weight (g) recorded on a sample of 12 fish are given below:

S. No	1	2	3	4	5	6	7	8	9	10	11	12
Length (cm)	17.6	19	27.2	20.2	18.6	14.8	21.1	16.8	21.4	13.2	23.7	24.6
Weight (g)	25	32	110	42	30	10	45	20	48	8	75	82

Fit length-weight relationship.

Solution

$\overline{X} = 1.2892, \overline{Y} = 1.2892$

$$b = \frac{n\sum XY - \left(\sum X\right)\left(\sum Y\right)}{n\sum X^2 - \left(\sum X\right)^2}$$

$$b = \frac{(12)(24.03) - (15.4701)(18.3722)}{(12)(20.0347) - (15.4701)^2}$$

$$b = \frac{4.1450}{1.0924}$$

b = 3.7944

$A = \overline{Y} - b\overline{X}$

A = 1.5310-4.8917

A = -3.3607

$s_y^2 = 0.12015$, $s_x^2 = 0.12015$

Variance of deviations from regression is given by

$$S^2 = \left(\frac{n-1}{n-2}\right)(s_y^2 - b^2 s_x^2)$$

$$S^2 = \left(\frac{11}{10}\right)(0.00101)$$

$S^2 = 0.0011$

a = Antilog $(A+S^2/2)$

a = Antilog (-3.3607 + (0.0011)/2)

a = 0.0004

The length-eight relationship is therefore given by,

$W = 0.0004\ L^{3.7944}$

22. Estimation of Marine Fish Landings

Sampling Design Developed by ICAR-CMFRI

- **Purpose:** To estimate the total marine fish landings in India.
- **Sampling Design:** Stratified Multistage Random Sampling.
- **Stratification:** Over space and time. Each maritime state is divided into several zones based on geographical considerations and fishing practices.

Stratified Multistage Random Sampling

Sampling for First-Stage Units

- Nine landing centers are randomly selected from each zone to record fish landings.
- A month is divided into three groups of 10 consecutive days.
- From the first 10-days group, a day is randomly selected to fall within the first five days.
- Then, six consecutive days starting from the selected day are considered and grouped into three clusters of two consecutive days each.
- From the second and third groups of 10 days, three clusters of two days each are systematically chosen with a sampling interval of 10 days.
- These nine clusters are allocated to the nine selected landing centers.
- Data are collected from 12 to 18 hours on the first day of observation and

from 6 to 12 hours on the next day.

- Data on night landings are collected through inquiry covering the period from 18 hours of the first day to 6 hours of the next day.
- This forms a landing center day and serves as the first-stage sampling unit.

Sampling for Second-Stage Units

- On the day of observation at the selected landing center, if the total number of fishing boats landing their catches is 10 or less, data on all the boats are collected.
- If the number of fishing boats exceeds 10, a sample of boats is selected in a predetermined manner.
- Fishing boats constitute the second-stage units on which data on species-wise catch, effort, craft, and gear are recorded.

Sampling for Third-Stage Units

- At the third stage, samples of commercially important species are taken from the selected second-stage units for biological observations.

Estimation of Total Landings

- Based on the data collected from the selected fishing boats, the total landings for the landing center day are estimated.
- From these estimates, monthly estimates for each year on a zonal, district, and state basis are calculated along with the corresponding sampling error.

Example-1

At a particular landing centre, data on fish landing were recorded during a month (30 days) on 10 days selected by systematic sample with a sampling interval of 3 days. Below are given the landings (in tons) of the 10 sampled days. Give an estimate of total fish landing of the centre for the months and find the approximate standard error

Days	1	2	3	4	5	6	7	8	9	10	Total
Landings(t)	3.5	4.5	3.2	3	4.2	4	3.5	5	4	3.3	38.2

Solution

N = 30, n = 10, K = 3

$$\sum_{i=1}^{10} \frac{Yi}{n}$$

= (3.5 + 4.5 + 3.2 + 3 + 4.2 + 4 + 3.5 + 5 + 4 + 3.3) −10

= 3.821

Total, $\hat{Y} = N \times \hat{Y}$

= 30 x 3.82

$\hat{Y}$ = 144.6

Variance of ($\bar{Y}$)

$$V(\bar{Y}) = \frac{N.N(K-1)}{2nK(n-1)} \left[\sum n\,(Yi+1-Yi)^2\right]$$

$$= \frac{(30)2(3-1)}{2(10)(3)(10-1)}\left[(Y_2 - Y_1))2 + (Y_3 - Y_2)2 + ..(Y_{10} - Y_9)2\right]$$

= 90(2) [1 + 169 + 1.4 + 4 + 0.04 + 0.025 + 2.25 +1 + 0.49] / 540

V(y) = 27.33

SE(y) = √V(y) = √27.33 = 5.227

SE($\hat{Y}$) = N√V(y) = 30 (5.227)

SE($\hat{Y}$) = 156.23

Example-2

Fishes are landed at a certain landing center throughout the year. 20 days were randomly selected from 365 days of a year and fish landed (in tonnes) during these days were recorded which are given below.

30,42,25,32,48,32,40,28,30,20,18,31,15,28,25,30,35,40,22,20.

Estimate,

- ❑ Average fish landing per day and total fish landing during the year
- ❑ Also find the S.E of total fish landing and average fish landing.

Solution

N = 365 days

- ❑ Average fish landing per day

$$\bar{y} = \sum \frac{yi}{N}$$

= 30+42+25+32+48+32+40+28+30+20+18+31+15+28+25+30+35+40+22+ 20/ 20

= 591/20

y = 29.551

Total fish yield of the year, $\hat{y} = N\bar{y}$

= 365 × 29.55

= 10785.75 tonnes

- ❑ Standard error of total fish landing and average fish landing

$$V(Y) = N^2 x\, V(\overline{Y})$$

$$= (365)^2 \text{ x } (\frac{1}{n} - \frac{1}{N})\, s^2$$

$$= (365)^2 \text{ x } (\frac{1}{20} - \frac{1}{365}) s^2$$

$$s^2 = \frac{1}{n-1} \sum_{i=1}^{n} (y_i - \bar{y})^2$$

$$= \frac{1}{19} \sum_{i=1}^{20} (y_i - n\bar{y})^2$$

$$= \frac{1}{19} (1388.95)$$

=73.1026

Hence , V(y)= 365^2 x (0.05 – 0.0027)73.10

= 460.642

$$\text{SE } (\hat{Y}) = \sqrt{V(\hat{Y})}$$

$$= \sqrt{460.642}$$

$$= 677.9$$

23. Estimation of Inland Fish Catch

- **1955-56:** Pilot investigation conducted by ICAR covered two districts of Hyderabad state.
- **April 1956:** The work was transferred from ICAR to the Directorate of National Sample Survey (NSSO), and work was carried out in Orissa.
- **End of 1958:** Basic information on various inland fisheries resources, their relative importance, and fishing practices became available.
- **1960-61:** Indian Statistical Institute in Calcutta attempted to evolve sampling methodology, but due to field problems, no estimation was attempted.
- **1962-63:** NSSO covered three districts of Orissa (Cuttack, Sambalpur, and Mayurbhanj).
- **1962-63:** CIFRI conducted work in Hoogly, West Bengal, attempting to estimate the area and catch from ponds.
- **1973-75:** NSSO covered West Bengal, Tamil Nadu, and Andhra Pradesh to estimate catch from impounded water and rivers using the inquiry method, but the estimates were not satisfactory.
- **1978-81:** IASRI & CIFRI conducted a pilot survey in West Bengal using inquiry and physical observation methods but couldn't cover all water resources due to limited manpower.
- **1984:** CIFRI conducted surveys in 8 states to evolve standardized methodology for collecting inland fisheries statistics in India.

- ☆ 156 - Research assessment survey works, 56 - catch assessment survey works.
- ☆ CIFRI formulated an estimation procedure for different ecological environments in inland fisheries (Bulletin Number 58).

Estimation Procedures of Inland Fish Catch

The estimation procedures involve the following steps:

i. **Sampling Frame**: The district is considered as the basic unit for estimation. The sampling frame is defined as the list of fishers and their active fishing day.

ii. **Stratification**: The population is stratified by water body type (rivers, reservoirs, lakes, etc.), season, and fishing intensity.

iii. **Catch Estimation**: The total catch is estimated by summing the catches of all fishers on their active fishing days. For rivers and estuaries, landing centers are used as the sampling units. For reservoirs and lakes, a stratified multistage sampling design is used

iv. **Variance Estimation**: Calculate the variance of the catch estimate using the formula for variance of a proportion.

Problems in Estimating Inland Fish Catch

- ❑ The diverse nature of inland fisheries resources has caused problems in evolving standard methodology for estimating fish catch.
- ❑ Important resources include rivers, canals, reservoirs, ponds, tanks, estuaries, beels, and oxbow lakes.
- ❑ Methodology evolved for one resource may not apply to another.
- ❑ Fish landing places are scattered, making access difficult.
- ❑ Harvesting operations in culture fisheries are staggered, and the fish catch reaching markets is after taking care of household requirements.
- ❑ An allowance for poaching also needs to be made while estimating fish catch.

Glossary

Alternative Hypothesis: A hypothesis that contradicts the null hypothesis, suggesting that there is a significant difference or effect.

ANOVA (Analysis of Variance): It's a statistical method used to compare the means of three or more groups to see if there are statistically significant differences between them. It partitions the total variation into different sources.

Autocorrelation: Autocorrelation measures the correlation between observations at different time points in a time series. It indicates whether there is a pattern or relationship between consecutive observations.

Binomial Distribution: A binomial distribution describes the probability of a certain number of successes in a fixed number of independent trials. It's characterized by two parameters: the number of trials and the probability of success.

Census: A complete enumeration of all individuals or items in a population.

Central Limit Theorem: The central limit theorem states that the sampling distribution of the sample mean approaches a normal distribution as the sample size increases, regardless of the shape of the population distribution.

Chi-Square Test: It's a statistical test used to determine whether there is a significant association between two categorical variables. It compares the observed frequencies to the expected frequencies.

Coefficient of Determination (R-squared): R-squared is a measure of how well the independent variables explain the variability of the dependent variable. It represents the proportion of variance in the dependent variable that is predictable from the independent variables.

Conditional Probability: The probability of an event occurring given that another event has already occurred.

Confidence Interval: It's a range of values calculated from sample data that is likely to contain the true population parameter with a certain level of confidence (e.g., 95% confidence interval).

Confounding Variable: A confounding variable is a variable that influences both the independent variable and the dependent variable, leading to a spurious association between them.

Correlation: It measures the strength and direction of the linear relationship between two variables. The correlation coefficient ranges from -1 to +1, where -1 indicates a perfect negative correlation, +1 indicates a perfect positive correlation, and 0 indicates no correlation.

Covariance: Covariance measures the degree to which two random variables change together. It indicates the direction of the linear relationship between variables.

Data: Facts, statistics, or information collected for analysis or reference.

Degrees of Freedom: Degrees of freedom represent the number of independent values or parameters that can vary in a statistical analysis. It's often used in calculating statistical tests.

Dependent Event: An event whose outcome is influenced by a previous event.

Descriptive Statistics: Methods used to describe and summarize the features of a dataset.

Diagrams: Visual representations of data or relationships, such as graphs, charts, or plots.

Differential Statistics: The study of changes in statistical measures over time or between groups.

Exclusive Class Interval: A class interval that includes its lower endpoint but not its upper endpoint.

Finite Population: A population with a limited or countable number of individuals or items.

F-Test: An F-test is a statistical test used to compare the variances of two populations. It's often used in analysis of variance (ANOVA) to test the equality of variances between groups.

Geometric Mean: A measure of central tendency calculated as the nth root of the product of a set of numbers, where n is the number of values.

Harmonic Mean: A measure of central tendency calculated as the reciprocal of the arithmetic mean of the reciprocals of a set of numbers.

Hypothesis Testing: It involves making a hypothesis about a population parameter and then collecting sample data to determine whether there's enough evidence to support or reject the hypothesis.

Inclusive Class Interval: A class interval that includes both of its endpoints.

Independent Event: An event whose outcome is not influenced by any previous events.

Inferential Statistics: Methods used to draw conclusions and make predictions about a population based on sample data.

Infinite Population: A population with an unlimited or uncountable number of individuals or items.

Interquartile Range (IQR): The interquartile range is the range between the first quartile (25th percentile) and the third quartile (75th percentile) of a dataset. It provides a measure of the spread of the middle 50% of the data.

Kurtosis: Kurtosis measures the "tailedness" of the probability distribution of a real-valued random variable. It indicates whether the distribution is more peaked or flat compared to the normal distribution.

Mean: It's calculated by adding up all the numbers in a dataset and then dividing by the total number of values. It represents the central value or average of the data.

Median: It's the middle value of a dataset when the numbers are arranged in ascending or descending order. If there's an even number of values, it's the average of the two middle numbers.

Mode: The mode is the value that appears most frequently in a dataset. A dataset may have one mode (unimodal), two modes (bimodal), or more than two modes (multimodal).

Multicollinearity: Multicollinearity occurs when independent variables in a regression model are highly correlated with each other. It can cause instability in the estimation of regression coefficients.

Nonparametric Statistics: It's a branch of statistics that doesn't require assumptions about the population distribution. Nonparametric methods are used when data don't meet the assumptions of parametric tests.

Normal Distribution: Also known as the Gaussian distribution, it's a symmetric bell-shaped curve characterized by its mean and standard deviation. Many natural phenomena tend to follow this distribution.

Null Hypothesis: A hypothesis that suggests there is no significant difference or effect.

Outlier: An outlier is an observation that lies an abnormal distance from other values in a dataset. It can significantly affect statistical analyses and should be investigated further.

Population: The population refers to the entire group of individuals, items, or data points that researchers are interested in studying and making inferences about.

Primary Data: Data collected firsthand by the researcher for a specific purpose.

Probability Distribution: A probability distribution describes the probabilities of different outcomes of a random phenomenon. It provides a mathematical description of the likelihood of each possible outcome.

Probability: Probability is a measure of the likelihood of an event occurring, ranging from 0 (impossible event) to 1 (certain event). It quantifies uncertainty in various situations.

Probability: The likelihood of a particular event occurring, often expressed as a number between 0 and 1.

p-value: The p-value represents the probability of obtaining results as extreme as the observed results, assuming that the null hypothesis is true. A smaller p-value indicates stronger evidence against the null hypothesis.

Random Variable: It's a variable whose value is subject to variations due to chance. It can take on different values according to the outcome of a random phenomenon.

Regression Analysis: It's a statistical method used to model the relationship between one or more independent variables and a dependent variable. It helps in understanding how changes in the independent variables affect the dependent variable.

Rejection Region: The set of values for a test statistic that leads to the rejection of the null hypothesis in hypothesis testing.

Residual: A residual is the difference between an observed value and the predicted value in a regression analysis. It represents the error in prediction.

Sample Survey: A study that collects data from a subset of individuals or items selected from a population to make inferences about the population.

Sample: A sample is a subset of the population that is selected for study. It's used to make inferences about the population as a whole.

Sampling: The process of selecting a subset of individuals or items from a population to estimate characteristics of the entire population.

Secondary Data: Data that have been collected by someone else for a different purpose but are utilized by researchers for their own analysis.

Skewness: Skewness measures the asymmetry of the probability distribution of a real-valued random variable about its mean. It indicates whether the data is skewed to the left or right.

Standard Deviation: It measures the amount of variation or dispersion in a set of values. A low standard deviation indicates that the values are close to the mean, while a high standard deviation indicates that the values are spread out.

T-Test: A T-test is a statistical test used to determine if there is a significant difference between the means of two groups. It's commonly used when the sample size is small or the population standard deviation is unknown.

Type I Error: Type I error occurs when the null hypothesis is incorrectly rejected, indicating that there is a significant effect or difference when there isn't one in reality. It's also known as a false positive.

Type II Error: Type II error occurs when the null hypothesis is incorrectly accepted, indicating that there is no significant effect or difference when there actually is one. It's also known as a false negative.

Variance: It measures the average squared deviation of each data point from the mean. It provides a measure of the spread or dispersion of the data.

Z-Score: A Z-score represents the number of standard deviations a data point is from the mean. It's used to standardize values and compare them to a standard normal distribution.

References

Birathar, R. S. 2002. Fisheries Statistics, Second Edition.

Chambers, J.M., 2018. Graphical methods for data analysis. Chapman and Hall/CRC.

Frank, H. and Althoen, S.C., 1994. Statistics: Concepts and applications workbook.

Gupta, S., 2002. Research methodology and statistical techniques. Deep and Deep Publications.

Gupta, S.C. and Kapoor, V.K., 2020. Fundamentals of mathematical statistics. Sultan Chand & Sons.

Hannan, E.J. and Deistler, M., 2012. The statistical theory of linear systems. Society for Industrial and Applied Mathematics.

Hayslett, H.T., 2014. Statistics. Elsevier.

Hinton, P.R., 2014. Statistics explained. Routledge.

Hoel, P.G., 1954. Introduction to mathematical statistics. Introduction to mathematical statistics., (2nd Ed).

Krishnaiah, P.R. ed., 1980. A Hand Book of Statistics (Vol. 1). Motilal Banarsidass Publishe.

Narain, P., Kathuria, O.P., Sharma, V.K. and Prajneshu, P. eds., 1991. Recent advances in agricultural statistics research.

Ostle, B., 1963. Statistics in research. Statistics in research., (2nd Ed).

Papoulis, A., 1990. Probability and statistics. Prentice-Hall, Inc..

Rangaswamy, R., 1995. A text book of agricultural statistics. new age international.

Rugg, G., 2007. using Statistics. McGraw-Hill Education (UK).

United States. National Agricultural Statistics Service, 1936. Agricultural statistics. United States Department of Agriculture, National Agricultural Statistics Service.

Lupton, R., 1993. Statistics in theory and practice. Princeton University Press.

9 789359 192307

Printed at Repro India Ltd.